ROTARY DRILLING SERIES

Marine Riser Systems and Subsea Blowout Preventers

Unit V, Lesson 10
First Edition

Formerly Subsea Blowout Preventers and Marine Riser Systems

By Hugh McCrae

Published by

PETROLEUM EXTENSION SERVICE
The University of Texas at Austin
Continuing Education
Austin, Texas

in cooperation with

INTERNATIONAL ASSOCIATION
OF DRILLING CONTRACTORS
Houston, Texas

2003

Library of Congress Cataloging-in-Publication Data

McCrae, Hugh.
Marine risers and subsea blowout preventers / by Hugh McCrae.— 1st ed.
p. cm. — (Rotary drilling series ; unit 3, lesson 4)
ISBN 0-88698-188-3
1. Oil wells—Blowouts—Prevention. 2. Oil wells—Safety measures.
3. Offshore oil well drilling. I. Title. II. Series.
TN871.215 .M38 2003
622'.3382—dc21
00-009600

First Edition published 1968. Second Edition 1983.
Third Edition 2001. Third Impression 2008
Printed in the United States of America

Catalog no. 2.51010
ISBN 0-88698-188-3

No state tax funds were used to publish this book.
The University of Texas at Austin is an equal opportunity employer.

Contents

Figures

Tables

Foreword

It has been almost thirty years since PETEX first issued a manual on subsea blowout preventers and marine riser systems. Technological advances have created economically viable solutions to the complications of well-control methods that are created by subsea blowout preventer systems. Further, wells are being drilled in waters deeper than ever before, and subsea technology has made it possible. New methods and equipment have changed the face of offshore drilling to allow companies to drill into ultra-deep waters. Without new techniques and equipment, the oil industry would not easily be able to tap into the offshore reservoirs that reside in formations lying below the deep waters of the world.

Because of the growing development of deepwater oil and gas fields, employees aboard floating rigs must be familiar with the specific problems of well control and understand how blowout prevention equipment and riser systems communicate with wells on the seafloor. Described in this book are the many components of marine riser systems, diverters, tensioning systems, wellheads and casing, BOP stacks, choke manifolds, mud and gas equipment, well-control equipment, and hydraulic control systems.

The writers of this new edition kept in mind the goal of this book: to explain and illustrate technological advances in a way that can be easily absorbed. The book informs and addresses the logistical and rig challenges operators face when planning and drilling subsea wells.

This book is designed to help employees do a better and safer job while drilling from floating offshore drilling vessels that employ subsea blowout preventer stacks.

Acknowledgments

PETEX owes a debt of gratitude to two people for the completion of *Marine Risers Systems and Subsea Blowout Preventers.*

Thanks first go to freelance writer, Robin Beckwith, who willingly took on the difficult job of researching and putting together the initial text for this book when no one else could. The job required that Robin research an unfamiliar and technical topic in an industry that has changed dramatically since the last printing of this book.

Since 1994 Beckwith has been a ghostwriter and editor in the oil and gas industry, initially working with Sperry-Sun Drilling Services producing technical articles, newsletters, and corporate videos, and later as a special correspondent with American Oil & Gas Reporter. She has also provided articles for major oil and gas and service companies such as Exxon, BP, Schlumberger, and Halliburton. She has been most recently employed as associate editor for *Oil & Gas Advisory*, an online email resource providing up-to-the-minute information regarding issues facing the oil and gas industry.

Secondly, this book would not be complete without the invaluable effort of Hugh McCrae, who generously offered his time and talent. As subsea superintendent of Transocean Engineering and Construction in Aberdeen, Scotland, Hugh has over 26 years of experience as engineer on various semis and drill ships, and currently develops, implements, and teaches in-house subsea training courses around the world. Offering a wealth of knowledge and information, Hugh revitalized this manuscript and provided PETEX with up-to-date drawings, illustrations, and photographs. Hugh's work on *Marine Risers Systems and Subsea Blowout Preventers* was extensive, and it is with his expertise that this book has been completed.

Finally, thanks go to Tom Thomas, of Transocean, Inc., for connecting PETEX with Hugh McCrae and for reviewing the final draft.

Units of Measurement

Throughout the world, two systems of measurement dominate: the English system and the metric system. Today, the United States is almost the only country that employs the English system.

The English system uses the pound as the unit of weight, the foot as the unit of length, and the gallon as the unit of capacity. In the English system, for example, 1 foot equals 12 inches, 1 yard equals 36 inches, and 1 mile equals 5,280 feet or 1,760 yards.

The metric system uses the gram as the unit of weight, the metre as the unit of length, and the litre as the unit of capacity. In the metric system, for example, 1 metre equals 10 decimetres, 100 centimetres, or 1,000 millimetres. A kilometre equals 1,000 metres. The metric system, unlike the English system, uses a base of 10; thus, it is easy to convert from one unit to another. To convert from one unit to another in the English system, you must memorize or look up the values.

In the late 1970s, the Eleventh General Conference on Weights and Measures described and adopted the Système International (SI) d'Unités. Conference participants based the SI system on the metric system and designed it as an international standard of measurement.

The *Rotary Drilling Series* gives both English and SI units. And because the SI system employs the British spelling of many of the terms, the book follows those spelling rules as well. The unit of length, for example, is *metre*, not *meter*. (Note, however, that the unit of weight is *gram*, not *gramme*.)

To aid U.S. readers in making and understanding the conversion to the SI system, we include the following table.

English-Units-to-SI-Units Conversion Factors

Quantity or Property	English Units	Multiply English Units By	To Obtain These SI Units
Length, depth, or height	inches (in.)	25.4	millimetres (mm)
		2.54	centimetres (cm)
	feet (ft)	0.3048	metres (m)
	yards (yd)	0.9144	metres (m)
	miles (mi)	1609.344	metres (m)
		1.61	kilometres (km)
Hole and pipe diameters, bit size	inches (in.)	25.4	millimetres (mm)
Drilling rate	feet per hour (ft/h)	0.3048	metres per hour (m/h)
Weight on bit	pounds (lb)	0.445	decanewtons (dN)
Nozzle size	32nds of an inch	0.8	millimetres (mm)
Volume	barrels (bbl)	0.159	cubic metres (m^3)
		159	litres (L)
	gallons per stroke (gal/stroke)	0.00379	cubic metres per stroke (m^3/stroke)
	ounces (oz)	29.57	millilitres (mL)
	cubic inches ($in.^3$)	16.387	cubic centimetres (cm^3)
	cubic feet (ft^3)	28.3169	litres (L)
		0.0283	cubic metres (m^3)
	quarts (qt)	0.9464	litres (L)
	gallons (gal)	3.7854	litres (L)
	gallons (gal)	0.00379	cubic metres (m^3)
	pounds per barrel (lb/bbl)	2.895	kilograms per cubic metre (kg/m^3)
	barrels per ton (bbl/tn)	0.175	cubic metres per tonne (m^3/t)
Pump output and flow rate	gallons per minute (gpm)	0.00379	cubic metres per minute (m^3/min)
	gallons per hour (gph)	0.00379	cubic metres per hour (m^3/h)
	barrels per stroke (bbl/stroke)	0.159	cubic metres per stroke (m^3/stroke)
	barrels per minute (bbl/min)	0.159	cubic metres per minute (m^3/min)
Pressure	pounds per square inch (psi)	6.895	kilopascals (kPa)
		0.006895	megapascals (MPa)
Temperature	degrees Fahrenheit (°F)	$\frac{°F - 32}{1.8}$	degrees Celsius (°C)
Thermal gradient	1°F per 60 feet	—	1°C per 33 metres
Mass (weight)	ounces (oz)	28.35	grams (g)
	pounds (lb)	453.59	grams (g)
		0.4536	kilograms (kg)
	tons (tn)	0.9072	tonnes (t)
	pounds per foot (lb/ft)	1.488	kilograms per metre (kg/m)
Mud weight	pounds per gallon (ppg)	119.82	kilograms per cubic metre (kg/m^3)
	pounds per cubic foot (lb/ft^3)	16.0	kilograms per cubic metre (kg/m^3)
Pressure gradient	pounds per square inch per foot (psi/ft)	22.621	kilopascals per metre (kPa/m)
Funnel viscosity	seconds per quart (s/qt)	1.057	seconds per litre (s/L)
Yield point	pounds per 100 square feet (lb/100 ft^2)	0.48	pascals (Pa)
Gel strength	pounds per 100 square feet (lb/100 ft^2)	0.48	pascals (Pa)
Filter cake thickness	32nds of an inch	0.8	millimetres (mm)
Power	horsepower (hp)	0.7	kilowatts (kW)
Area	square inches ($in.^2$)	6.45	square centimetres (cm^2)
	square feet (ft^2)	0.0929	square metres (m^2)
	square yards (yd^2)	0.8361	square metres (m^2)
	square miles (mi^2)	2.59	square kilometres (km^2)
	acre (ac)	0.40	hectare (ha)
Drilling line wear	ton-miles (tn•mi)	14.317	megajoules (MJ)
		1.459	tonne-kilometres (t•km)
Torque	foot-pounds (ft•lb)	1.3558	newton metres (N•m)

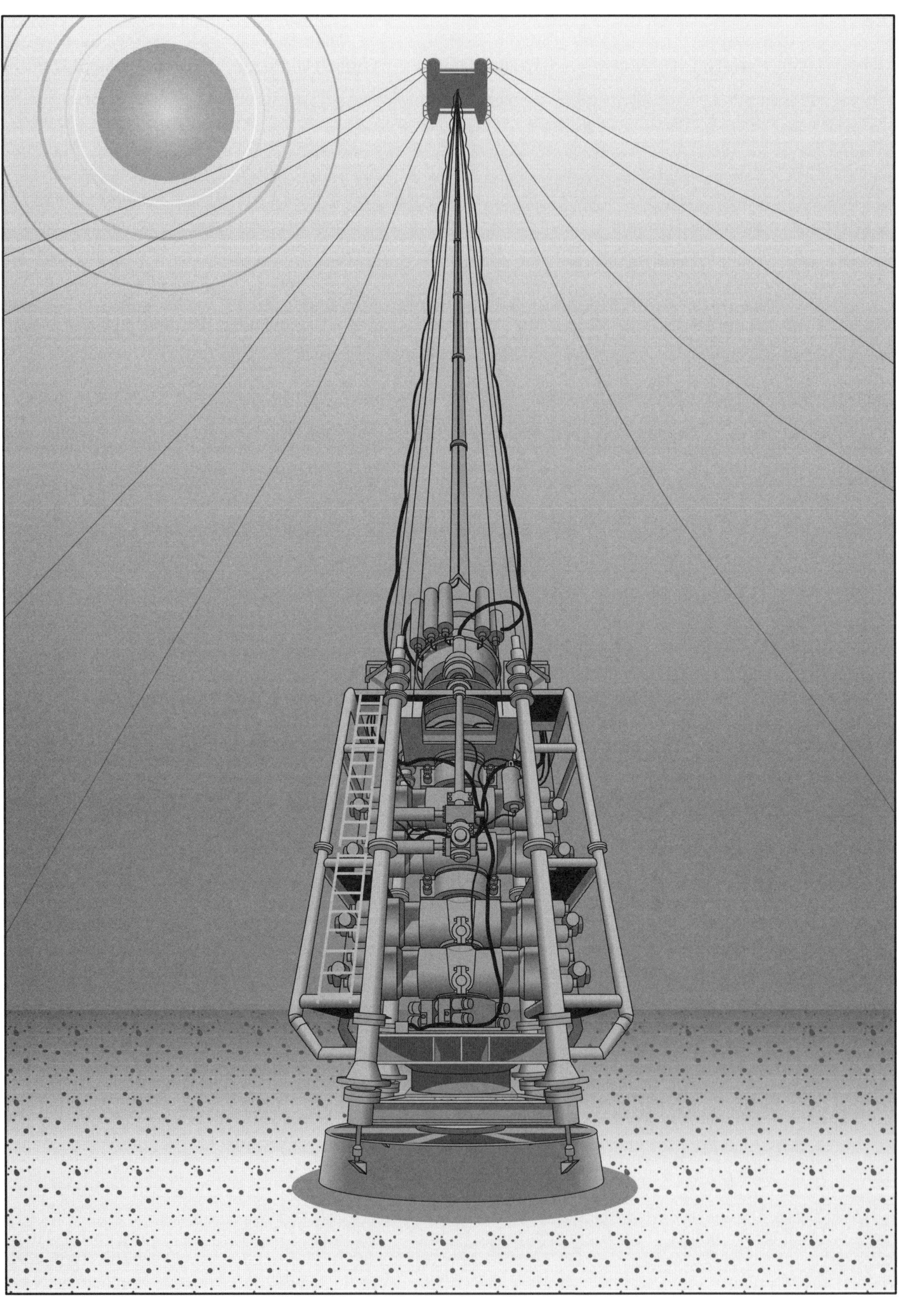

Introduction

During normal drilling operations, the column of drilling fluid—the mud—in the well creates hydrostatic pressure to counterbalance formation pressures. As long as the rig crew maintains the mud's hydrostatic pressure at a value greater than the formation pressure, formation fluids cannot flow into the well. In short, the mud column is the primary means of well control.

In the English, or conventional, system of measurement, where mud weight is in pounds per gallon (ppg), hole depth is in feet (ft), and pressure is in pounds per square inch (psi), the following equation can be used to calculate the hydrostatic pressure created by the mud column at the bottom of the well.

$$HP = 0.052 \times MW \times TVD \qquad \text{(Eq. 1)}$$

where

HP = hydrostatic pressure, psi
MW = mud weight, ppg
TVD = true vertical depth, ft.

The number 0.052 is a constant—that is, its value does not change as long as the other units of measure in the equation, such as ft, ppg, and psi, do not change. The constant 0.052 is derived from the pressure in psi created by a 1-ft high column of fluid that has a weight of 1 ppg.

As an example of determining hydrostatic pressure in psi, suppose a well is full of mud that weighs 12.0 ppg and that the well's *TVD* is 12,500 ft. What is the hydrostatic pressure at the bottom of the well? Using equation 1—

$$\begin{aligned} HP &= 0.052 \times 12.0 \times 12{,}500 \\ &= 0.624 \times 12{,}500 \\ HP &= 7{,}800 \text{ psi.} \end{aligned}$$

In the SI system of measurement, which is based on the metric system, mud weight is in kilograms per cubic metre (kg/m^3), depth is in metres (m), and pressure is in kilopascals (kPa). In SI units, the following equation can be used to calculate hydrostatic pressure at the bottom of the well.

$$HP = 0.0098 \times MW \times TVD \qquad \text{(Eq. 2)}$$

where

HP = hydrostatic pressure, kPa
MW = mud weight, kg/m^3
TVD = true vertical depth, m.

The constant 0.0098 is derived from the fact that 1 pascal (Pa) equals 1 newton (N) of force applied to 1 square metre (m^2) and that a cube measuring 1 m on each side contains 1 m^3. One N exerts 9.8 kg of force on the bottom of 1 m^3. To obtain the pressure in kPa, divide the force on the cube by 1,000. This calculation gives 0.0098.

As an example of determining hydrostatic pressure in kPa, suppose a well is full of mud that weighs 1,437.8 kg/m^3 and that the well's *TVD* is 3,810 m. What is the hydrostatic pressure at the bottom of the well? Using equation 2—

$$HP = 0.0098 \times 1{,}437.8 \times 3{,}810$$
$$= 14.090 \times 3{,}810$$
$$HP = 53{,}683 \text{ kPa.}$$

Regardless of the units employed, the hydrostatic pressure the mud column creates can be altered in two ways: (1) by changing the mud weight, or (2) by changing the mud column's height. For example, suppose the well in the previous example is drilled to 13,000 ft (3,962 m) with the original 12.0-ppg (1,437.8-kg/m^3) mud weight. In this case—

$$HP = 0.052 \times 12.0 \times 13{,}000$$
$$= 0.624 \times 13{,}000$$
$$HP = 8{,}112 \text{ psi.}$$

Or, in SI units—

$$HP = 0.0098 \times 1{,}437.8 \times 3{,}962$$
$$= 14.909 \times 3{,}962$$
$$HP = 55{,}825 \text{ kPa.}$$

By increasing the depth of the well, but keeping it full of 12.0-ppg (1,437.8-k/m^3) mud, the hydrostatic pressure at the bottom of a 13,000-ft (3,962-m) well increases from 7,800 to 8,112 psi (53,683 to 56,796 kPa).

Similarly, hydrostatic pressure can be changed by increasing the mud weight from 12.0 (1,437.8) to 12.5 ppg (1,497.8 kg/m^3), while the well's *TVD* remains unchanged at 12,500 ft (3,810 m). For example, in English units—

$$
\begin{aligned}
HP &= 0.052 \times 12.5 \times 12{,}500 \\
&= 0.65 \times 12{,}500 \\
HP &= 8{,}125 \text{ psi.}
\end{aligned}
$$

In this case, increasing the mud weight from 12.0 to 12.5 ppg increases hydrostatic pressure at *TVD* from 7,800 to 8,125 psi.

In SI units—

$$
\begin{aligned}
HP &= 0.0098 \times 1{,}497.8 \times 3{,}810 \\
&= 14.678 \times 3{,}810 \\
HP &= 55{,}923 \text{ kPa.}
\end{aligned}
$$

Increasing the mud weight from 1,437.8 to 1,497.8 kg/m^3 increases hydrostatic pressure at *TVD* from 53,683 to 55,923 kPa.

If formation pressure becomes greater than the mud column's hydrostatic pressure, the well may *kick*. A kick is the uncontrolled flow of formation fluids into the well. These formation fluids may be water, gas, or oil, alone or in combination. The rig crew must contain the intruded fluids to prevent the kick from becoming an uncontrolled *blowout*. A blowout is the uncontrolled flow of formation fluids to the atmosphere or into another formation exposed to the wellbore. Blowouts can be hazardous to rig personnel, rig equipment, and the environment.

Formation pressure can become greater than hydrostatic pressure under different circumstances. For example, it can occur when crew members allow the mud column's height to drop when tripping pipe from the hole. As the crew removes pipe, they must replace the metal volume they remove with an equal amount of mud; otherwise, the mud level will drop. If it drops enough, the length of the column of mud does not develop hydrostatic pressure equal to or above formation pressure. Consequently, the well kicks.

Formation pressure can exceed hydrostatic pressure when the hole penetrates an abnormally pressured formation. If the mud weight is established for normal formation pressure, and the bit encounters an abnormally pressured zone, the abnormally high formation pressure overcomes hydrostatic pressure. Thus, the well kicks.

Formation pressure can exceed hydrostatic pressure if crew members fail to degas *gas-cut mud*. Gas-cut mud occurs when small pockets of gas mix with the mud while drilling and reduce the mud's hydrostatic pressure.

If the rig crew swabs formation fluids into the hole while pulling pipe, formation pressure can become greater than hydrostatic pressure. Swabbing occurs when the upward motion of the drill string causes a negative pressure at the bottom of the well. The pressure reduction caused by swabbing allows formation fluids to flow into the well. If the crew swabs in enough formation fluids to reduce the mud's hydrostatic pressure below formation pressure, the well will kick.

Most operators regard hydrostatic pressure as the primary means of well control; they usually consider well-control equipment as the secondary means of well control. Crew members use the equipment to contain well kicks and prevent them from becoming blowouts. Various components make up well-control equipment, including ram preventers, annular preventers, choke and kill systems, marine riser systems, and diverter systems. In floating drilling operations, the blowout prevention (BOP) equipment is usually located on the ocean floor. The marine riser system extends the wellbore from the ocean floor to the drilling vessel on the surface. This manual describes the BOP equipment used in floating drilling applications.

To summarize—

- In the English system of measurement, where mud weight (*MW*) is in pounds per gallon and true vertical depth (*TVD*) is in feet, hydrostatic pressure (*HP*) in psi can be determined with the equation:

$$HP = 0.052 \times MW \times TVD.$$

- In the SI system of measurement, where mud weight (*MW*) is in kilograms per cubic metre and true vertical depth (*TVD*) is in metres, hydrostatic pressure (*HP*) in kilopascals can be determined with the equation:

$$HP = 0.0098 \times MW \times TVD.$$

- Hydrostatic pressure may be changed by changing the mud's weight or changing the height of the mud column.
- A kick can occur when the hydrostatic pressure developed by the column of mud in the well is less than formation pressure.
- A blowout is the uncontrolled flow of formation fluids into the atmosphere or into an underground formation.
- Formation pressure can exceed hydrostatic pressure when—
  1. the mud column's height is allowed to drop to too low a height when tripping pipe—that is, the crew fails to fill the hole to replace the volume removed by the drill string.
  2. the hole penetrates a formation with abnormally high pressure.
  3. the crew fails to degas gas-cut mud.
  4. swabbing occurs.

Marine Riser Systems

A marine riser system provides a fluid conduit to and from the wellbore—that is, it extends the wellbore from the subsea BOP to the drilling rig. It also supports auxiliary lines, such as high-pressure choke and kill lines, mud booster lines, and hydraulic conduits. Further, the marine riser system guides the drill stem and other tools from the drilling rig to the wellhead on the seabed. Finally, it provides a means of running and retrieving the BOP assembly from the surface to the wellhead on the seafloor.

Marine riser systems are critical equipment; therefore, if a system fails, catastrophic losses can result. Consequently, the overall design of the system is of paramount importance. Generally, system design begins with an assessment of expected operating conditions and an engineering analysis to establish such factors as tensile loads, bending stresses, maximum operational water depth, buoyancy requirements, surface tension, and vessel response to motion.

Additional factors that affect riser system design include—

- dynamic and axial loads while running and retrieving the riser system and BOP assembly;
- lateral forces from currents and vessel offset;
- cyclic forces from wave and vessel motion;
- vortex induced vibrations (VIVs);
- axial loads created by the weight of the riser system itself, the weight of the drilling fluid inside the riser, and the additional weight of freestanding pipe within the riser;
- axial tension from the tensioning system at the surface;
- dimensional requirements such as outside diameter and wall thickness of the main riser pipe;

- outside diameter (OD) and pressure rating of the choke and kill lines;
- OD and pressure rating of the mud booster line;
- OD and pressure rating of the hydraulic conduit line;
- method and makeup time of joint connections;
- storage and handling requirements; and
- lifetime operating and maintenance costs.

A typical marine riser system consists of surface tensioning equipment; a surface diverter system; a telescopic, or slip, joint; individual riser joints; high-pressure choke and kill lines; hydraulic conduit lines; a mud booster line; a riser fill-up valve; a termination spool; an instrumented riser joint; a lower marine riser package (LMRP), which consists of flexible piping for the choke and kill, hydraulic conduit, and mud booster lines; a flex joint; one or two annular BOPs; and a hydraulic connector for connecting the LMRP to the BOP stack.

Riser Pipe

Figure 1 shows the components that make up a riser assembly from the telescopic joint at the surface to the flex joint on the LMRP. A series of individual riser joints make up the assembly between the

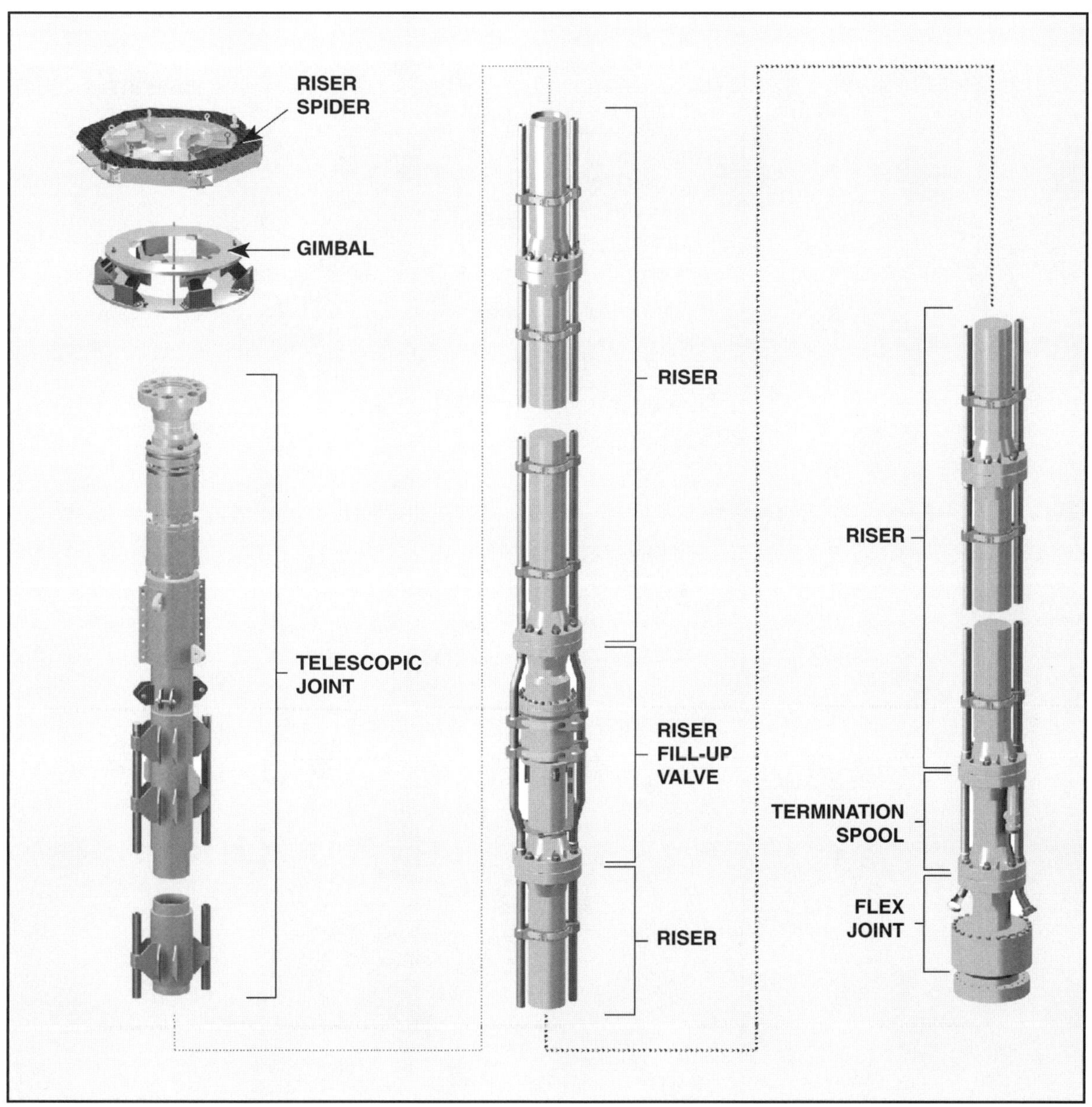

Figure 1. Riser pipe and connectors (Courtesy of Cameron)

telescopic joint and the flex joint. A riser joint (fig. 2) is a large-diameter, seamless or welded high-strength pipe that has connectors at each end. Special clamps hold the auxiliary lines to the pipe.

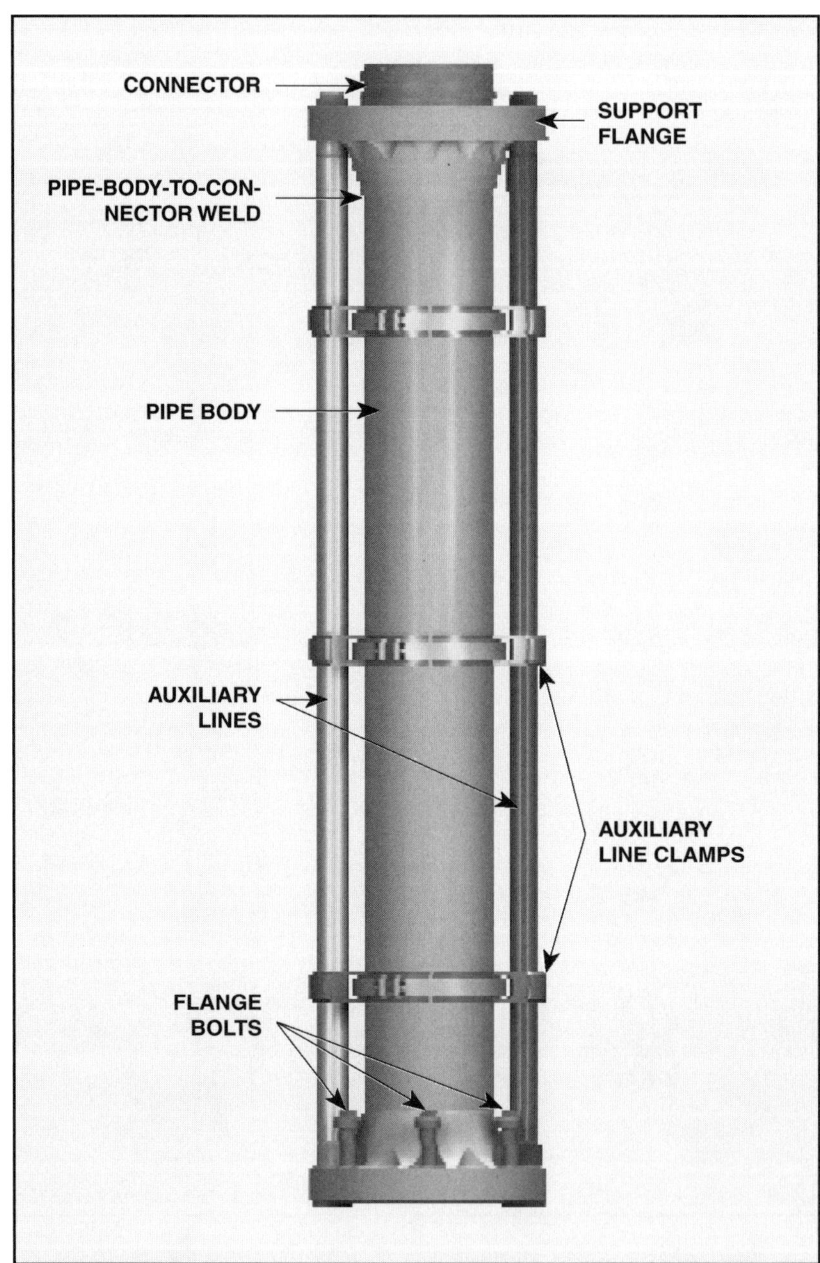

Figure 2. Various components that make up a riser assembly (Courtesy of Cameron)

When the rig deploys the BOP assembly and riser system, crew members connect the first riser joint to the LMRP's flex joint, usually on top of the termination spool. They then connect subsequent joints on the drill floor and lower them into the water. The string of riser joints represents the principal component of the riser system.

As crew members lower the riser joints into the water, they place the upper end of each joint onto a spider assembly at the rotary table. A support flange on the upper end of the riser joint's connector goes into the spider assembly. The spider supports both static and dynamic loads as the next joint is being connected. The support flange also supports the auxiliary lines and the upward force created by buoyant modules (if they are used) as they enter the water below the rotary table.

Riser pipe is classified by its outside diameter, wall thickness, and material grade. A minimum amount of internal clearance is allowed to permit the running of drilling assemblies, casing, full-bore wellhead tools and casing hangers. Table 1 shows the combinations of BOP inside diameter (ID) and riser OD that are commonly used.

Table 1
Riser ODs Depend on BOP IDs

BOP ID, in. (mm)	Riser OD, in. (mm)
13⅝ (346)	16 (406.4)
16¾ (425.5)	18⅝ (473.1)
18¾ (476.3)	21 (533.4)
21¼ (539.8)	24 (609.6)

Manufacturers offer riser joints in several lengths. Typical lengths are 50, 65, 70, 75 and 90 ft (15.24, 19.81, 21.34, 22.86, and 27.43 m). Lengths of 70 ft (21.34 m) and longer are most often used in deepwater applications. For shallower water operations, 50-ft (15.24-m) lengths are common because they are easier to handle than longer joints. Manufacturers also make riser *pup joints*, which are short lengths of riser used to *space out* the riser assembly as it is run in various water depths. (To space out means to install one or more riser pup joints to obtain the necessary overall riser length from the BOP stack to the telescopic joint.) Rigs employ pup joints 5, 10, 15, 20, 25, 30, 35, and 40 ft (1.5, 3.0, 4.6, 6.1, 7.6, 9.1, 10.7, and 12.2 m) long to space out the riser assembly. When the crew selects pup joints for space-out, they should be sure the telescopic joint is positioned in its midstroke position. With the telescopic joint at midstroke, it ensures that the joint provides its maximum available stroke.

Riser manufacturers use steel that has good fatigue characteristics and is formulated to allow connectors to be welded on the riser without creating problems. Most manufacturers use X80-grade steel, which has a *yield strength* of 80,000 psi or 551,600 kPa. (Yield strength is a measure of the amount of pressure inside or outside the riser that is required to permanently distort it.)

Riser wall thickness increases as *tensile load* requirements increase with water depth. (Tensile load is longitudinal stress placed on a riser by its weight. As the weight increases, longitudinal stresses increase.) Commonly, manufacturers offer risers with wall thickness of ½, ⅝, 11⁄16, ¾, 1, and 1¼ in. (12.7, 15.9, 17.5, 19.1, 25.4, and 31.8 mm). In cases where wall thickness impairs the ability to run full-bore tools through the riser, the riser OD has to be increased. For example, a 21-in. (533.4-mm) OD riser may have a wall thickness of 1¼ in. (31.8 mm) and an ID of 18½ in. (469.9 mm). In this case, crew members cannot run full-bore 18¾-in. (476.3-mm) tools through it. Thus, a 21¼- or 21½-in. (539.8- or 546.1-mm) OD riser should be selected.

The internal pressure rating of the main riser pipe should be at least equal to the working pressure of the surface diverter system, plus the maximum difference in hydrostatic pressure between the drilling fluid inside the riser and the seawater outside. The difference between hydrostatic pressure inside and seawater pressure outside may be as much as 0.440 psi/ft (9.953 kPa/m) of water depth, based on a maximum mud weight of 17 ppg (2,036.9 kg/m^3). This differential is more than 1,000 psi (6,895 kPa) for a water depth of 2,500 ft (762 m). High internal pressures are only one of the many challenges deep water drilling presents, even if drilling fluids lighter than 17 ppg (2,036.9 kg/m^3) may be in use.

Rig engineers must also consider riser collapse resistance in deepwater drilling. Riser collapse can occur when internal drilling fluid pressure drops below external seawater pressure. Such a condition can occur if a lost circulation zone is drilled into. In this case, the mud drains out of the riser and into the zone of loss. Also, mud can fall out of the riser pipe during an emergency disconnect.

Crew members can install an automatic flooding valve (fig. 3) in a deepwater riser system to prevent collapse of the riser when accidental evacuation of the mud in the riser occurs. The valve is incorporated into a riser joint and senses any increase in differential pressure. Once the differential pressure reaches a preset value, the valve automatically opens an outer sleeve, which allows seawater to fill the riser. Besides automatic operation, a rig crew member can operate the valve remotely from the surface using a hydraulic control panel.

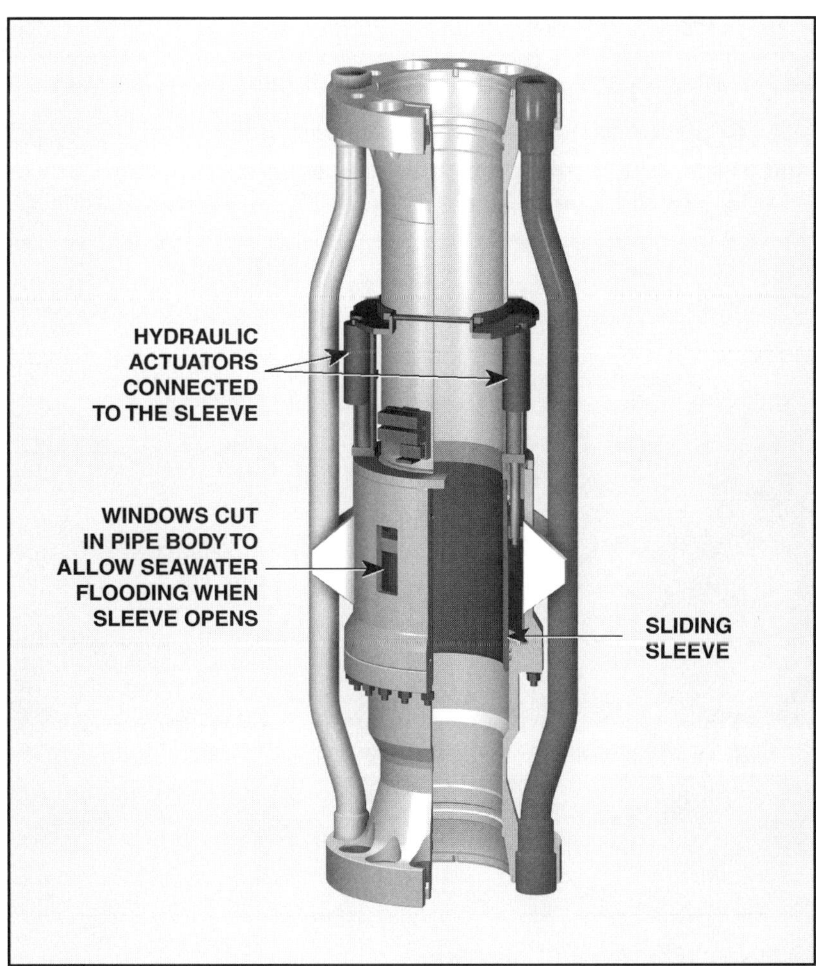

Figure 3. Automatic flooding valve (Courtesy of Cameron)

Riser Connectors

Because riser joint connectors operate under extreme conditions, manufacturers make them stronger than riser pipe. Riser connectors must resist separation at their maximum design load. Separation leads to failure. The connections are therefore preloaded to their design load rating.

Rigs use two main types of connector: the dog type and the bolted-flange type. One bolted-flange connector is the Cameron Load King connector (fig. 4). It is designed for drilling in water depths of 7,000 ft (2,100 m) or more and has a tensile load rating of 3.5 million pounds (lb) or 1,557,500 decanewtons (dN). This connector requires a relatively high torque on the flange bolts because they are part of the load path. Applying torque preloads the connection to withstand separating forces. Separating forces include those caused by wind, waves, and currents; vessel movement; and the load of the BOP assembly when it is run and retrieved. Crew members apply the required torque on the flange bolts with special, hydraulically operated, high-capacity torque wrenches. They must apply the manufacturer's recommended torque to ensure the correct preload in the connection. As stated earlier, any separation at the flanges can lead to failure.

Replaceable threaded inserts are installed in both the upper and lower flanges; thus, if a thread is damaged, the insert can be

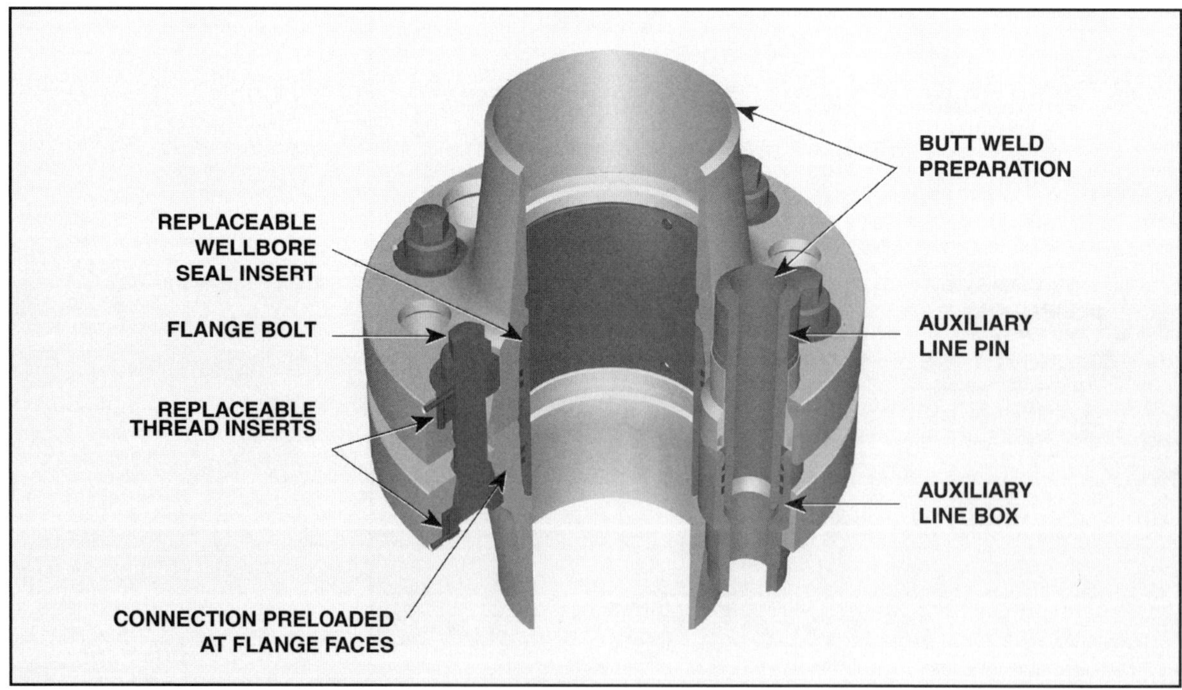

Figure 4. Cameron Load King connector (Courtesy of Cameron)

easily replaced. Crew members normally leave the bolts engaged in the top flange thread when disconnected to keep the bolts from being damaged or lost. Replaceable wellbore seal inserts with self-energizing lip seals are provided to seal the connection.

The Cameron RF flanged connector (fig. 5) is similar to the Load King connector, but has a tensile load rating of 2 million lb (890,000 dN).

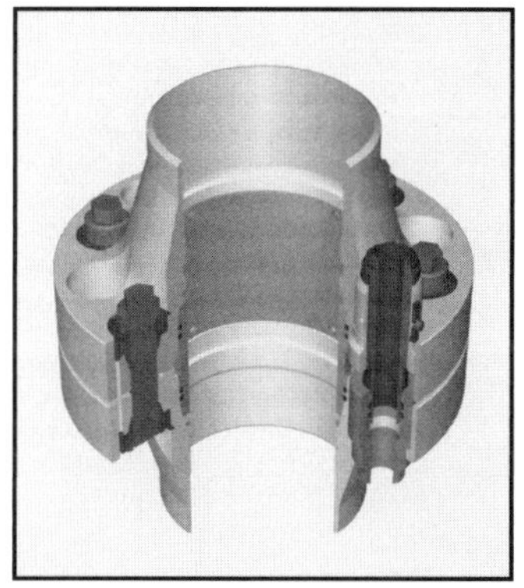

Figure 5. Cameron RF flanged connector (Courtesy of Cameron)

Dog-type connectors (fig. 6) have actuator screw assemblies in the box end. Using pneumatic torque wrenches, crew members tighten the actuator screws to make up the joints. As the screws go in, they cause tapered locking dogs to engage mating tapers in the pin end. The application of the correct torque to the actuator screw, combined with the taper geometry of the pin and locking dogs, preloads the connection to the design load rating. Spring-loaded locks prevent vibration from backing out the actuator screws. The actuator screws do not support any of the riser connection's load.

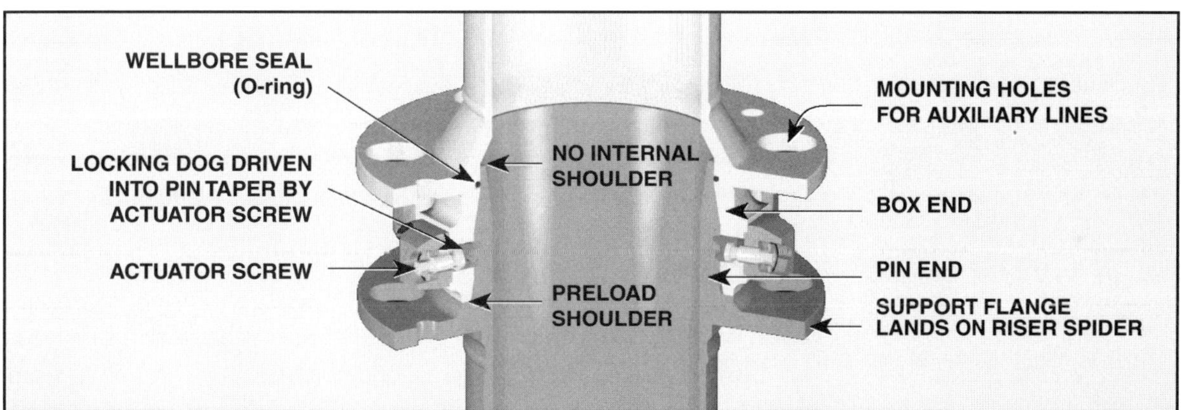

Figure 6. Dog-type riser connector (Courtesy of Shaffer)

Termination Spools

In most applications, the LMRP's riser adapter is installed directly to the top of the flex joint. A riser connector at the top of the adapter provides a place for the crew to make up the first riser joint. A mud booster line is connected to an opening—a *side entry*—in the wall of the riser adapter. The riser adapter adds height to the BOP-LMRP assembly; on some rigs, the cellar deck is not big enough to handle the additional height. Where a height restriction occurs, the side entry can be excluded from the riser adapter to significantly reduce its height

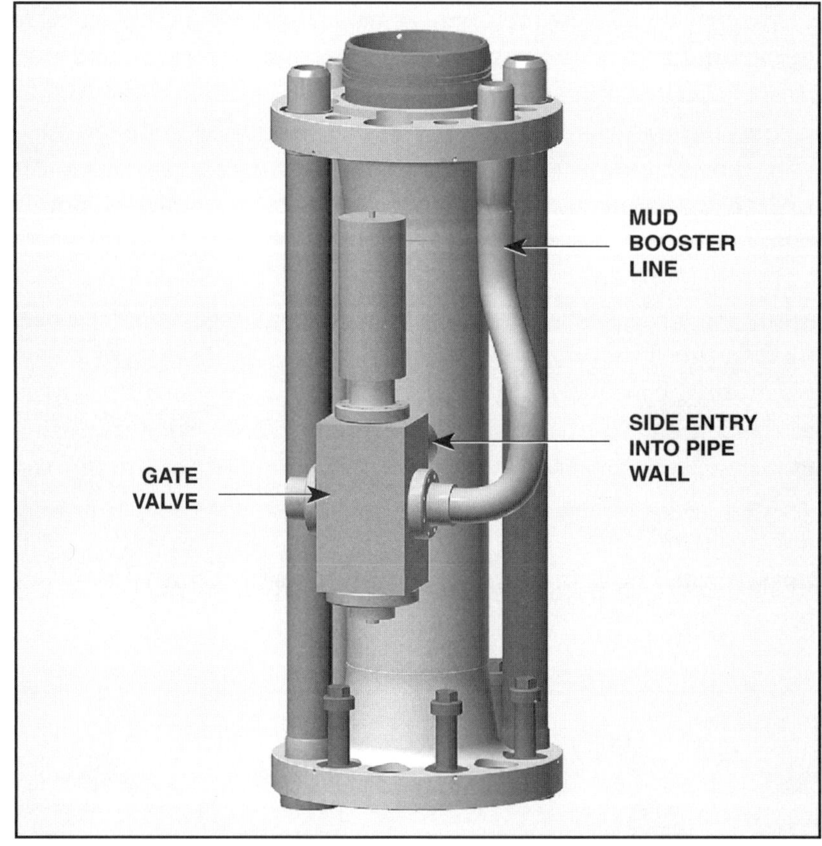

Figure 7. Termination joint, or spool (Courtesy of Cameron)

To overcome a height restriction, the rig can use a *termination joint*, or *spool*, which is merely a standard riser joint with a side entry to allow connection of the mud booster line (fig. 7.) When used, the termination spool is the first riser joint connected to the LMRP. Some spools feature an automatic check valve that allows mud to flow only down the circulation line and up the riser string. A hydraulically operated fail-safe gate valve can also be used (see fig. 7).

Telescopic Joint

The telescopic, or slip, joint is installed at the top of a marine riser system. A telescopic joint—

- compensates for vertical movement (heave) of the vessel,
- provides a means of connecting the diverter assembly to the riser,
- provides terminations for the riser auxiliary lines to flexible hoses at the drilling vessel, and
- provides attachment points for the riser tensioning system.

A telescopic joint (fig. 8) is made up of an outer barrel and an inner barrel. The outer barrel is attached to the top joint of the marine riser assembly. The inner barrel is attached to the diverter assembly's flex joint. Riser tensioning lines are attached either to a

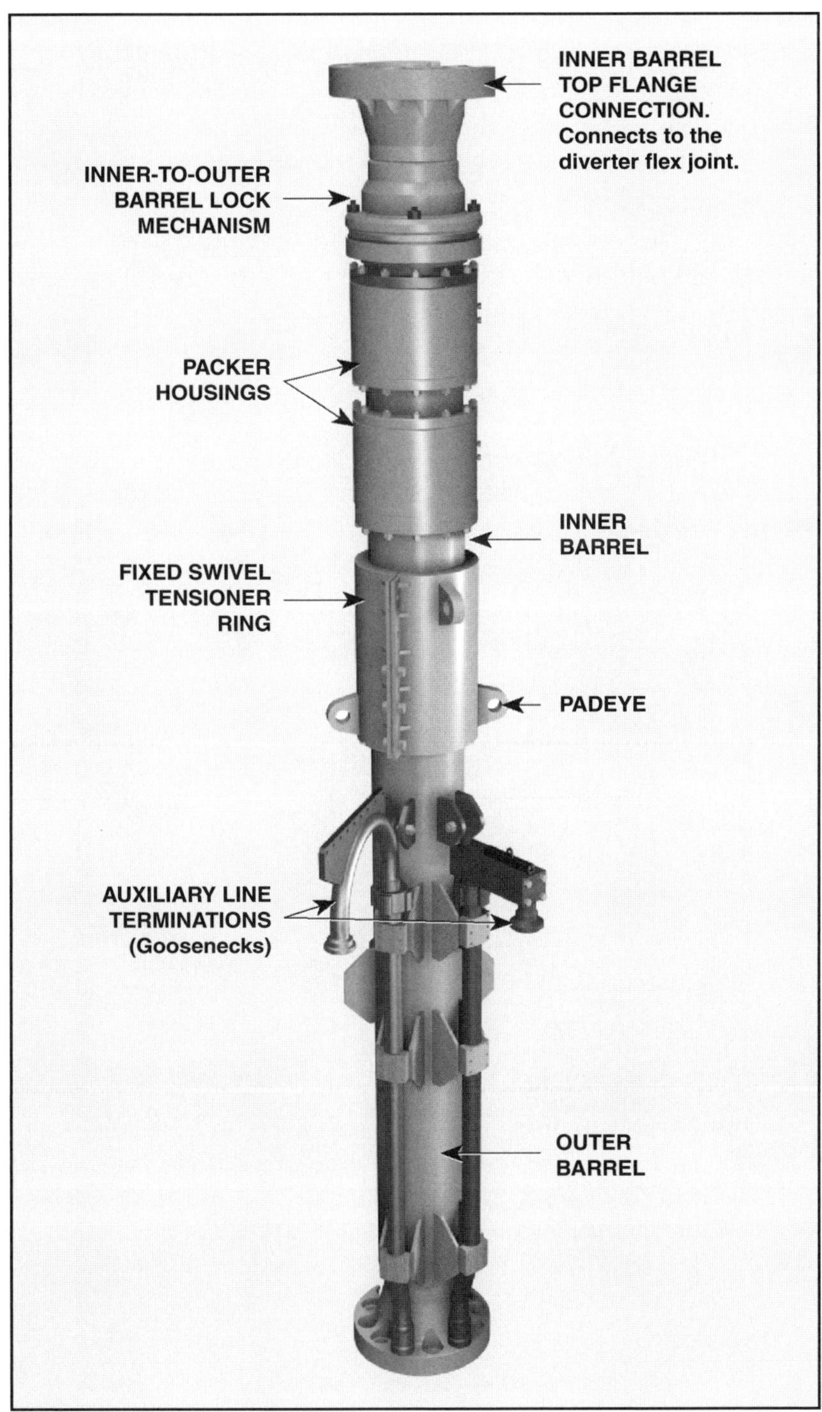

Figure 8. Telescopic joint (Courtesy of Cameron)

swivel tensioner ring or to a hydraulic tensioner ring on the outer barrel. A swivel tensioner ring is bolted to the outer barrel and can rotate around the outer barrel.

Once the telescopic joint is lowered below the rotary table, crew members install the riser tensioner lines. If the riser has a swivel tensioner ring, they manually install the tensioner lines to the padeyes on the ring after the joint is lowered through the rotary. Similarly, they have to manually remove the tensioner lines before the joint can be raised to the rig floor through the rotary table.

A hydraulic tensioner ring (fig. 9) allows crew members to run and retrieve the telescopic joint without removing the tensioner lines. The ring is hydraulically latched to the outer barrel as the joint is run, and hydraulically latched to the diverter housing as the joint is being retrieved.

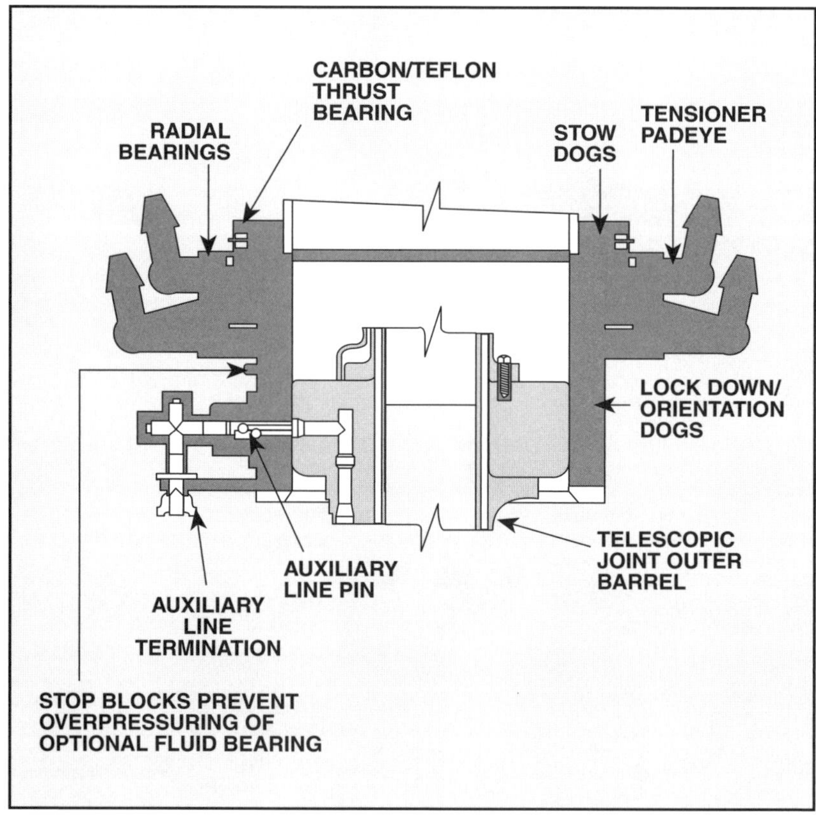

Figure 9. Hydraulic tensioner ring (Courtesy of ABB Vetco Gray)

The stroke length of the telescopic joint can be from 45 to 65 ft (about 15 to 20 m) depending on the rig's requirements. A 45-ft (15-m) stroke is normally sufficient for shallow water, where the environment is relatively mild. On the other hand, a 65-ft (20-m) stroke may be required for operations in deep water where the environment is harsh. Figure 10 shows a telescopic joint in operation.

Locking devices on the telescopic joint lock the inner barrel to the outer barrel in the fully retracted position. Fully retracted, the joint is at its shortest, which allows for easier handling when shipping and when being picked up on the rig or at a shore base. When running or retrieving the BOP assembly, the inner and outer barrels are always locked in the retracted position.

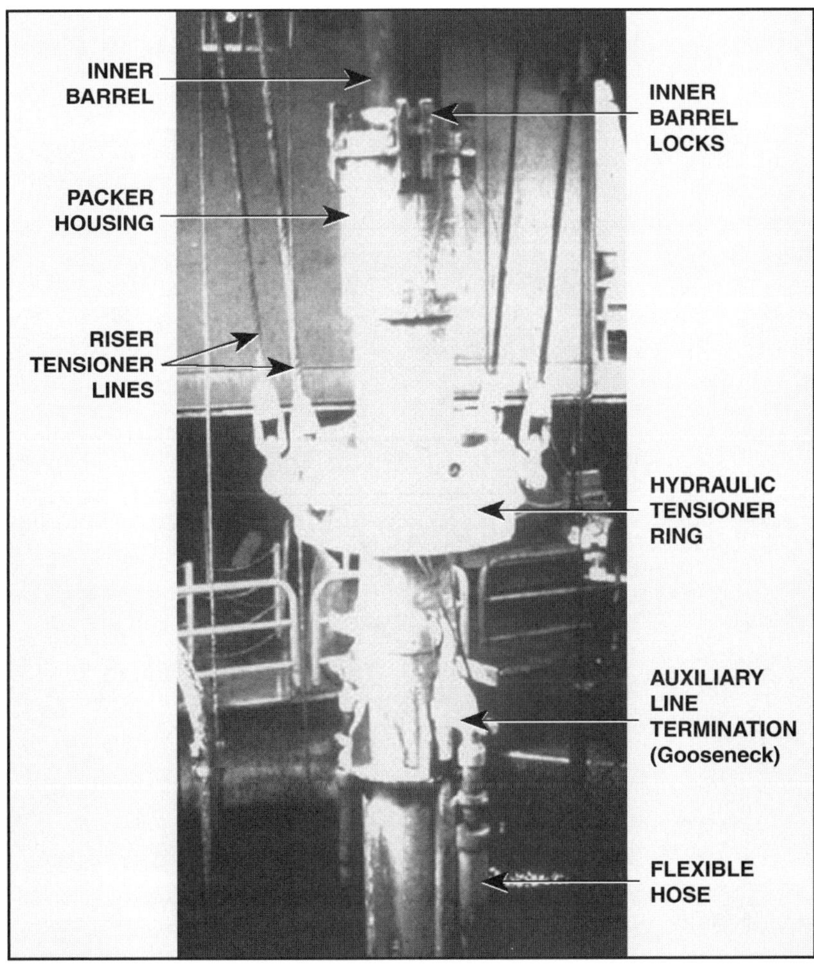

Figure 10. Telescopic joint with hydraulic tensioner ring (Courtesy of ABB Vetco Gray)

One or two housings on the outer barrel store pressure-energized radial packer seals (see fig. 8). A packer provides a mud seal between the outer barrel and inner barrel—that is, the packer keeps mud inside the inner barrel (fig. 11). Normally, regulated rig air pressure energizes the packer seals. Where two seals are provided, the lower one can be held in reserve, in an as-new condition, to be actuated only if it becomes necessary to divert a shallow gas flow.

On some two-seal systems, hydraulic pressure, rather than rig air, energizes the lower seal when diverting shallow gas kicks. Hydraulic pressure provides a much higher sealing pressure than rig air pressure. The packer seals are of the split type, which make it possible to change a worn upper seal while the telescopic joint is in place and drilling is going on. The lower packer is energized to maintain a seal while the upper seal is changed.

Water lubrication minimizes wear on the seals, which is caused by the constant motion of the inner barrel. To further reduce seal wear, the air-pressure regulator should be set to provide the least amount of pressure required to maintain a seal.

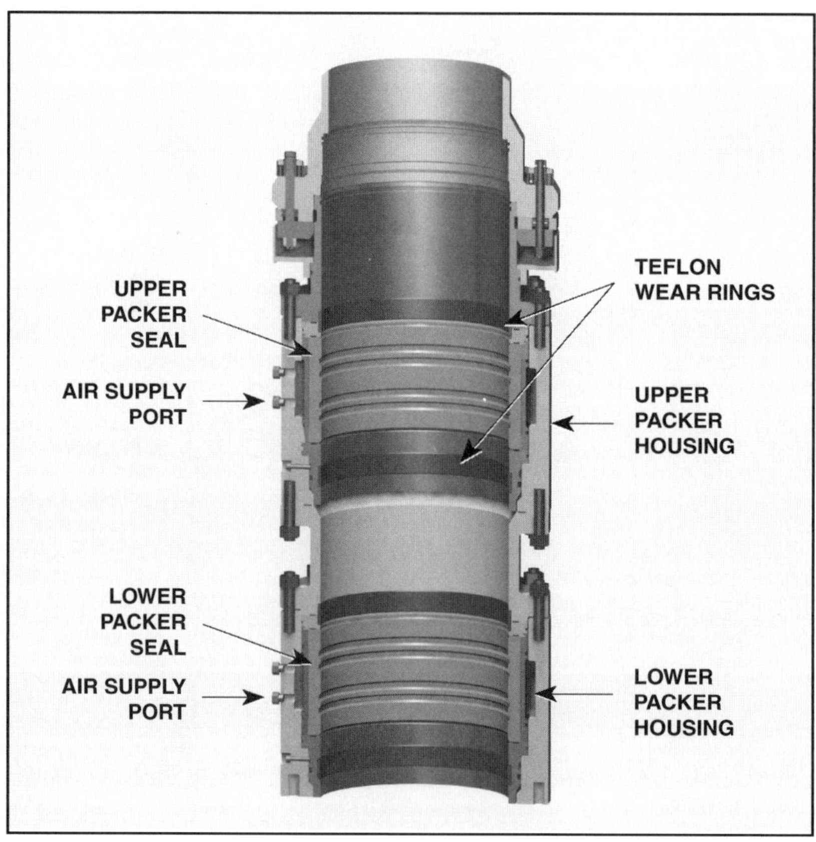

Figure 11. Packer seals in a telescopic joint (Courtesy of Cameron)

Flexible hoses (see fig. 10), which terminate on the floating rig's piping, are connected to goosenecks on the telescopic joint. The goosenecks are at the end of the riser-mounted auxiliary lines (fig. 12). Two types of goosenecks are available: a long sweep and a target type.

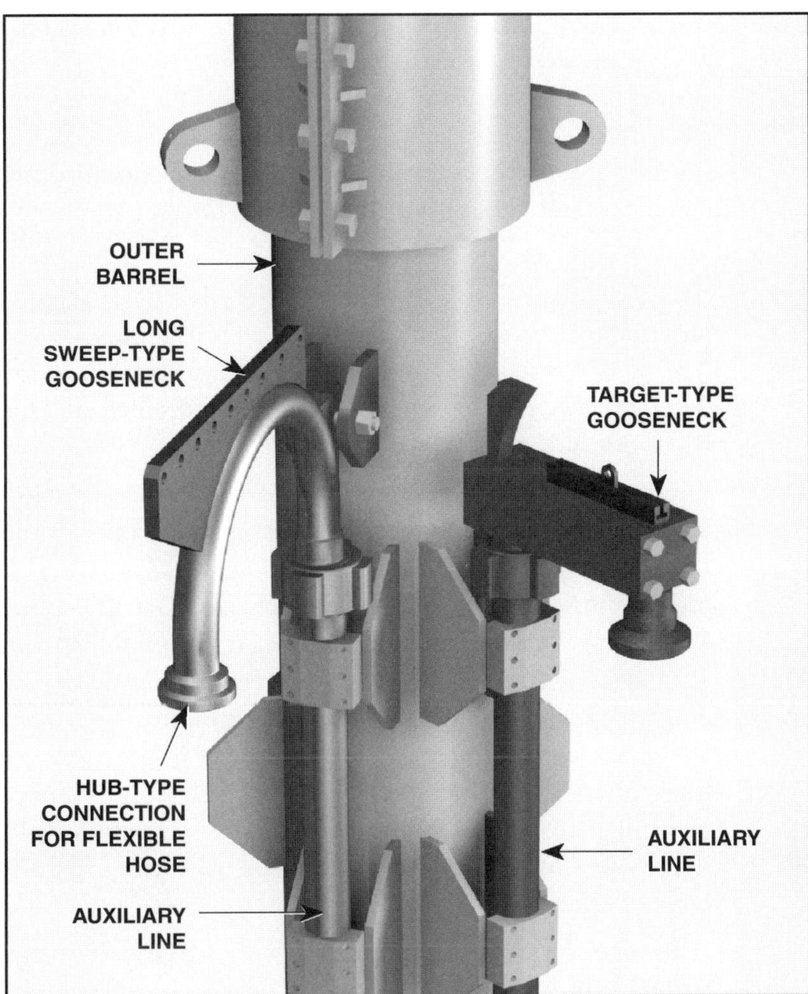

Figure 12. Gooseneck terminations (Courtesy of Cameron)

Auxiliary Lines

Auxiliary lines include choke and kill lines, mud booster lines, and hydraulic conduit lines. General guidelines to keep in mind about auxiliary lines include the following.

- To prevent accidental mismatching of auxiliary lines when the riser joints are being made up, the lines should be placed asymmetrically around the riser flange. Asymmetric placement ensures that the joints mate correctly. To prevent accidental overpressuring while testing, the test caps for the choke and kill lines are designed so that they cannot be installed on the mud booster line or hydraulic conduit lines. The mud booster line and hydraulic conduit lines cannot tolerate the high pressures required for choke and kill line testing.
- The choke and kill lines and mud booster line are constructed of metal that meets the requirements of the National Association of Corrosion Engineers (NACE) Specification MR-01-75, which covers steels intended for hydrogen sulfide (H_2S) service.
- The auxiliary line couplings must be able to seal against full pressure, while allowing for relative motion between the box and pin. Note that a phenomenon called *Poisson's effect* creates motion between the box and pin on the auxiliary line couplings. Poisson's effect is movement that results from structural compression caused by pressure exerted on the ends of the pins. Also, temperature differences between the fluids in the main riser and the fluids in the auxiliary lines can cause motion. Moreover, riser deflections impose bending loads. All relative motion can cause fatigue cracking of the support flange if an adequate gap is not provided between the support flange and the coupling.
- On some deepwater systems, the shoulders on the box and pins of the choke and kill lines, as well as the support flanges, share a portion of the main riser pipe's load.

Choke and Kill Lines

The riser-mounted, high-pressure choke and kill lines serve to extend the wellbore to the drilling rig when the subsea BOP is closed on a well kick. During well killing operations, well fluids normally return to the surface through the choke line. If required, mud can be pumped down the kill line.

Manufacturers make both lines from a suitable grade of steel, usually X52, X65, or X80, which have a yield of 52,000, 65,000, and 80,000 psi (358,540, 448,175, and 551,600 kPa) respectively. Also, they make them with a suitable OD and wall thickness to ensure that the lines have a pressure rating equal to the ram BOP's rating.

Their ID is generally 3 in. (76.2 mm). In deepwater applications, however, the lines' ID may be 4½ in. (114.3 mm) to help reduce pressure losses caused by friction as fluids travel inside the lines. (The larger the diameter of the lines are, the less pressure lost because of friction.)

Crew members can (1) circulate mud down one line and take returns up the other, (2) circulate down the drill stem and take returns up one line, or (3) circulate down the drill stem and take returns up both lines.

Shoulders and recesses on the riser's pin and box flanges hold the choke and kill lines on the riser joint. Also, clamps provide support for the choke and kill lines. The clamps are spaced at regular intervals on the riser joint. Some systems use a lock nut on one end of the line to allow the gap between the lines to be adjusted. The gap between the lines' shoulder and flange must be set to the correct amount. An incorrect gap can create excessive bending forces on the riser support flanges when pressure is applied. Using the correct number and properly spacing the clamps is also important to the pressure retaining ability of the lines. The clamps prevent the lines from excessive bending under pressure. Too much bending could result in high stress and the failure at the pin-and-box connection.

When crew members stab a riser connection, the pin-and-box connections of the choke and kill lines are stabbed simultaneously. The main riser connection, the line shoulders, and the riser flanges prevent separation of the pin and box when pressure is applied to the lines.

The choke and kill line boxes contain self-energizing lip seals, which form a pressure seal on the highly polished pin-end stab. These seals provide a pressure-tight seal from zero to maximum working pressure. The seal from zero to maximum working pressure is required because kicks can be of low or high pressure and the seal must be maintained at all pressures.

Riser handling tools have special test caps, which allow crew members to test the lines as they run the riser. They usually test the lines about every eight to ten joints while the joint is sitting in the riser spider.

Booster Lines

A mud booster line is sometimes provided on riser joints to increase the return velocity of the fluid column in the riser. The mud booster line is connected to a pump on the rig that enables the driller to circulate drilling fluid from the surface to the bottom of the riser. Circulating fluid in this manner helps move heavy cuttings up the riser.

Manufacturers make mud booster lines from a suitable grade of steel, usually X52 or X65, which have a yield of 52,000 and 65,000 psi (358,540 and 448,175 kPa) respectively. Booster lines have a suitable OD and wall thickness to provide a pressure rating equal to the mud circulating system rating. Their ID is generally 3 in. (76.2 mm).

Hydraulic Conduit Lines

For deepwater applications, the BOP control system's hydraulic fluid is supplied through riser-mounted conduit lines. Some systems have two conduit lines, one for each control pod, and some have just one line, which supplies both pods. A manifold system on the LMRP directs the fluid from the conduit line to the selected control pod.

Because the hydraulic control fluid must be clean for the control system to operate properly, these lines are normally manufactured from stainless steel to prevent rust from occurring and contaminating the fluid. The line pressure rating has to be the same as the BOP control system rating. Hydraulic conduit lines normally have a 2- or 3-in. (50.8- or 76.2-mm) ID. To prevent a corrosion cell from forming where stainless and regular steel components come into contact, galvanic protection is provided.

Handling Tools and Spiders

The rig crew uses riser handling tools to raise and lower the riser and telescopic joints. Handling tools have the same tensile load rating as the riser system and can support the full load of the riser and BOP assembly. A mechanical riser handling tool (fig. 13) duplicates a riser joint connection. It is made up to a riser joint just as though a riser joint was being connected. The upper section of the handling tool is a standard drill pipe tool joint that crew members latch the elevators onto to raise and lower riser joints.

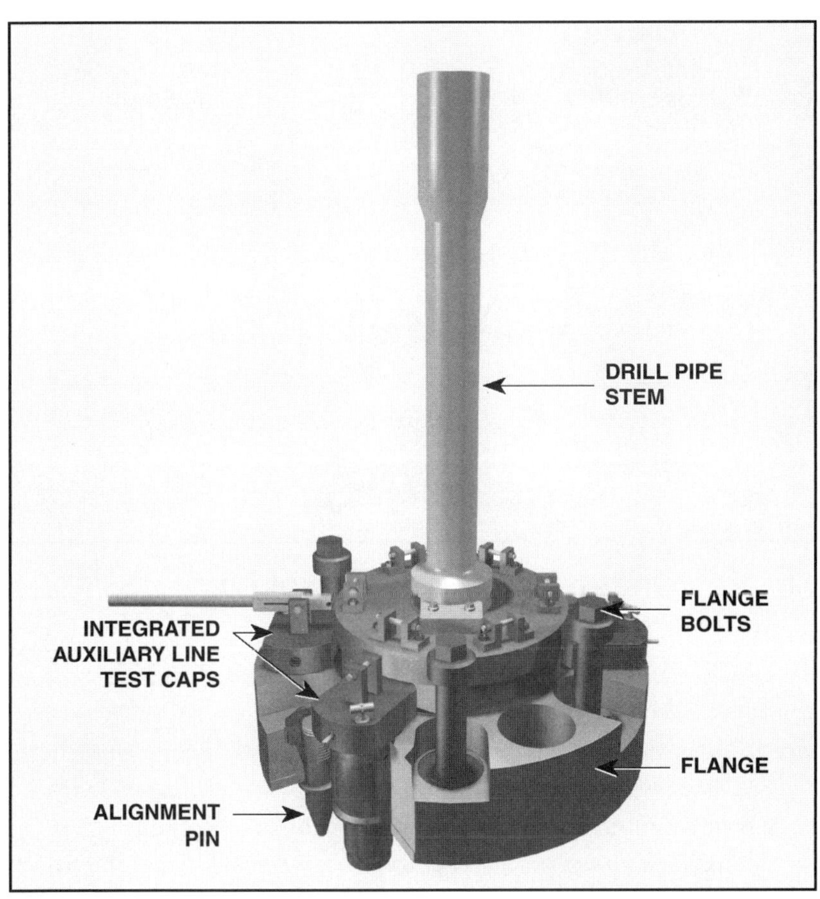

Figure 13. Mechanical riser handling tool (Courtesy of Cameron)

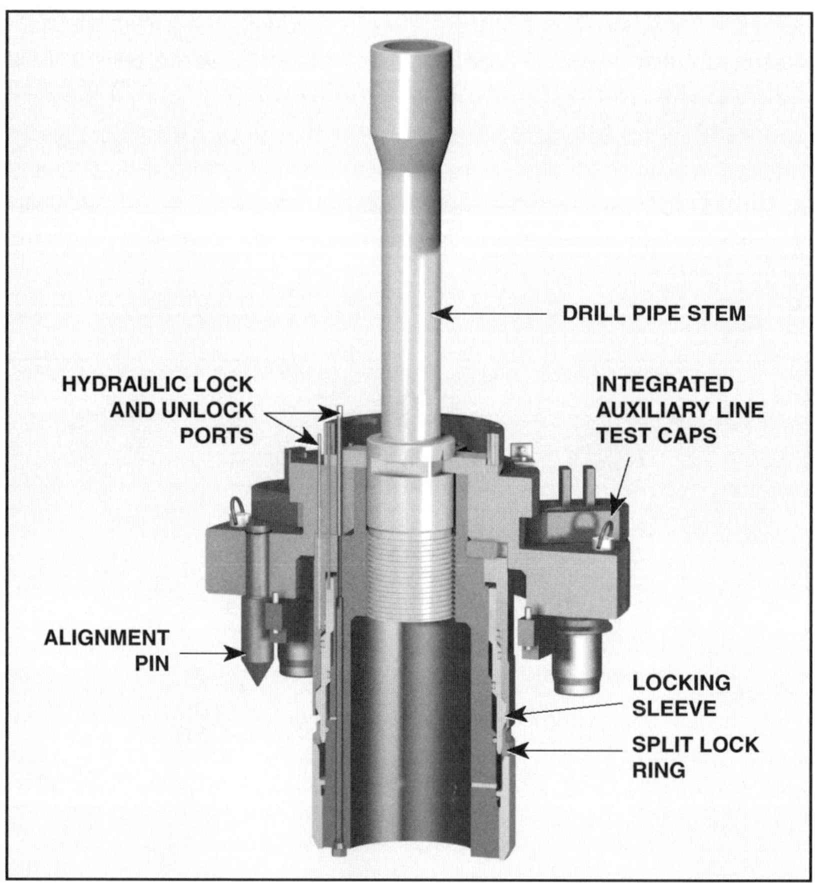

Figure 14. Hydraulic riser handling tool (Courtesy of Cameron)

A hydraulic riser handling tool (fig. 14) is similar to a mechanical tool, but it is operated hydraulically to reduce the time involved in making the connections. Hydraulic pressure forces a sleeve downward behind a split lock ring, which expands into a recess in the body of the riser connection.

Crew members place a *riser-support spider* (fig. 15) at the rotary table. The riser and BOP assembly rest on the riser-support spider as the crew picks up and adds each new joint to the string. The riser spider's ID is the same as the rotary table's ID, and the base plate is designed for the specific rotary table with which it is used.

The spider consists of a base plate and top plate with a series of support dogs, or arms, between them. The dogs support the riser joint under the riser joint's support flange. Hydraulic cylinders force the support dogs inward and under the riser joint support flange.

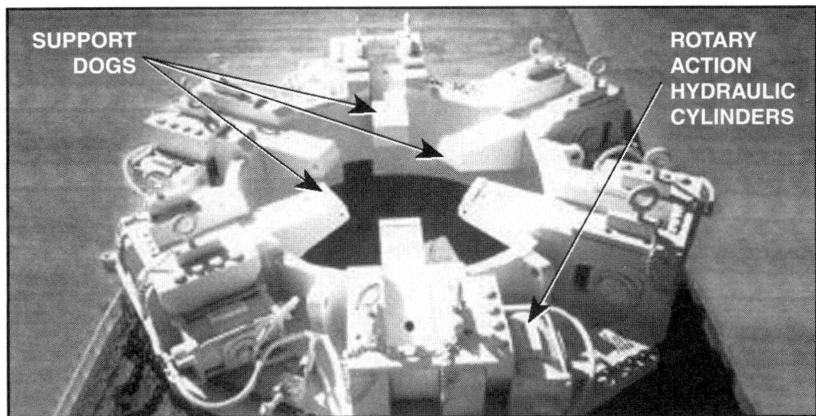

Figure 15. Riser-support spider (Courtesy of Cameron)

They are also retracted hydraulically to allow the riser joints to pass through. On some riser spiders, the support arms are engaged and disengaged manually.

In deepwater applications, a support gimbal (fig. 16) can be placed between the rotary table and the riser spider. The gimbal has hydraulic cylinders, or flex elements, that support the weight of the spider, the entire riser string, and BOP assembly. A support gimbal acts much like a flex joint, in that it compensates for riser offset caused by currents and maintains the riser flange square to the drill floor for making connections. It also cushions shock loads imparted on the system.

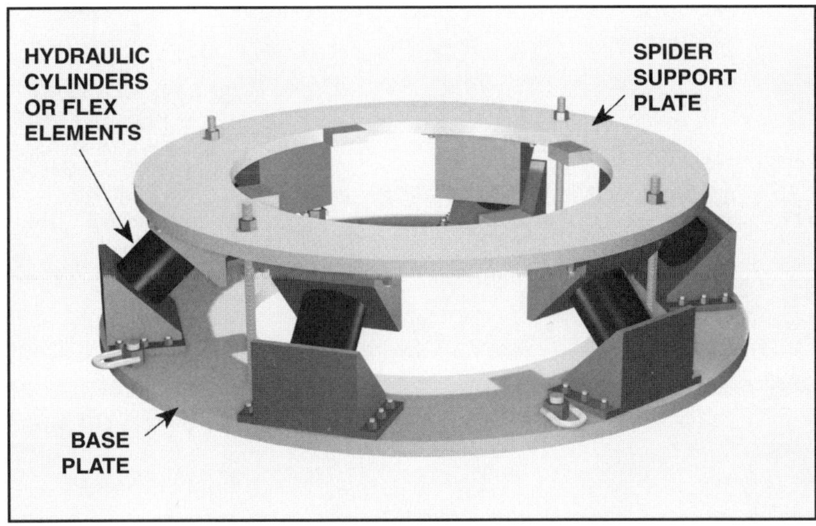

Figure 16. A support gimbal (Courtesy of Cameron)

Buoyant Riser Modules

In deepwater drilling—drilling in water depths beyond 2,000 ft (600 m)—the total submerged weight of the riser string must be kept as low as possible. Even heavy-duty riser tensioning equipment cannot handle the weight of a long riser string required for deepwater drilling. Reducing the weight of the riser string is achieved by adding buoyant modules, which are made from *syntactic foam*, to the outside of each riser joint. Syntactic foam is a thermoset resin with reinforced micro- and macrospheres. In other words, syntactic foam is plastic foam that is hardened by heat and is full of small and round air bubbles. The air bubbles are encased in a hard shell. The modules are generally 15 ft (4.6 m) long, are scalloped to fit around the auxiliary lines, and are retained to the riser by a strap and tensioner assembly placed around the circumference (fig. 17). Modules are supplied for use in various water depths, from 2,000 to 10,000 ft (600 to 3,000 m). Installed correctly, buoyant riser modules can reduce the effective weight of the riser string in water by 90 to 95 percent of its weight in air.

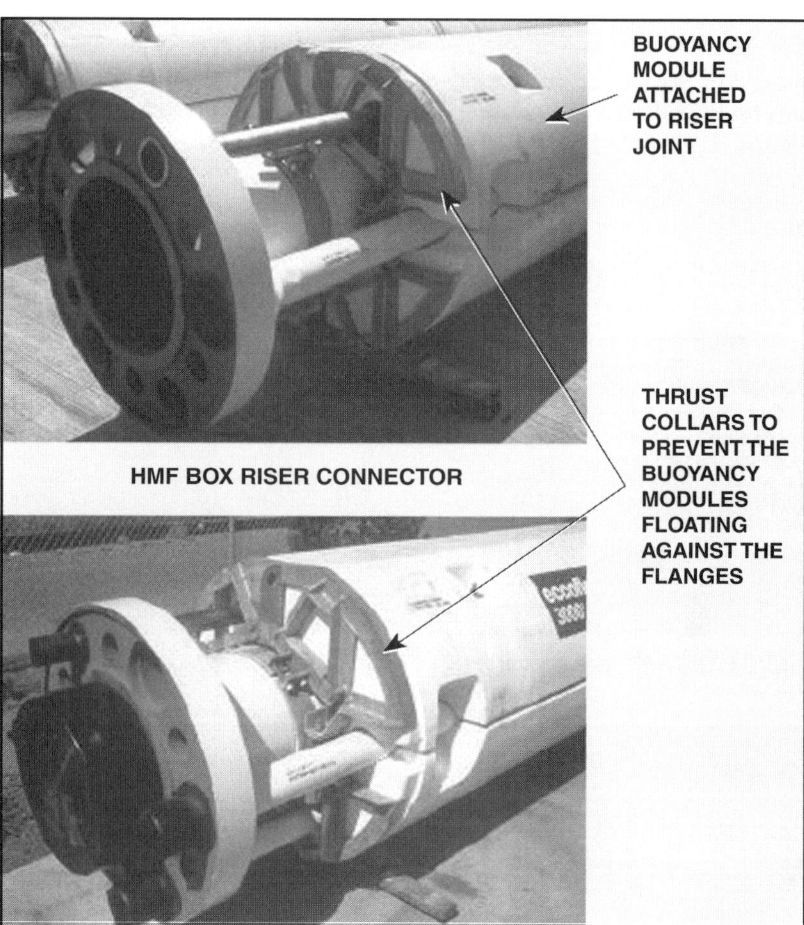

Figure 17. Buoyant modules attached to riser joints (Courtesy of ABB Vetco Gray)

Lower Riser Marine Package (LMRP)

The lower riser marine package, or LMRP, (fig. 18) is an assembly of several components, which include—

- a riser adapter,
- a flex or ball joint,
- an annular BOP,
- BOP control pods,
- a hydraulic connector,
- flexible choke and kill lines, and
- choke and kill stabs.

The LMRP is an interface between the riser system and the BOP stack that crew members can disconnect should the need arise.

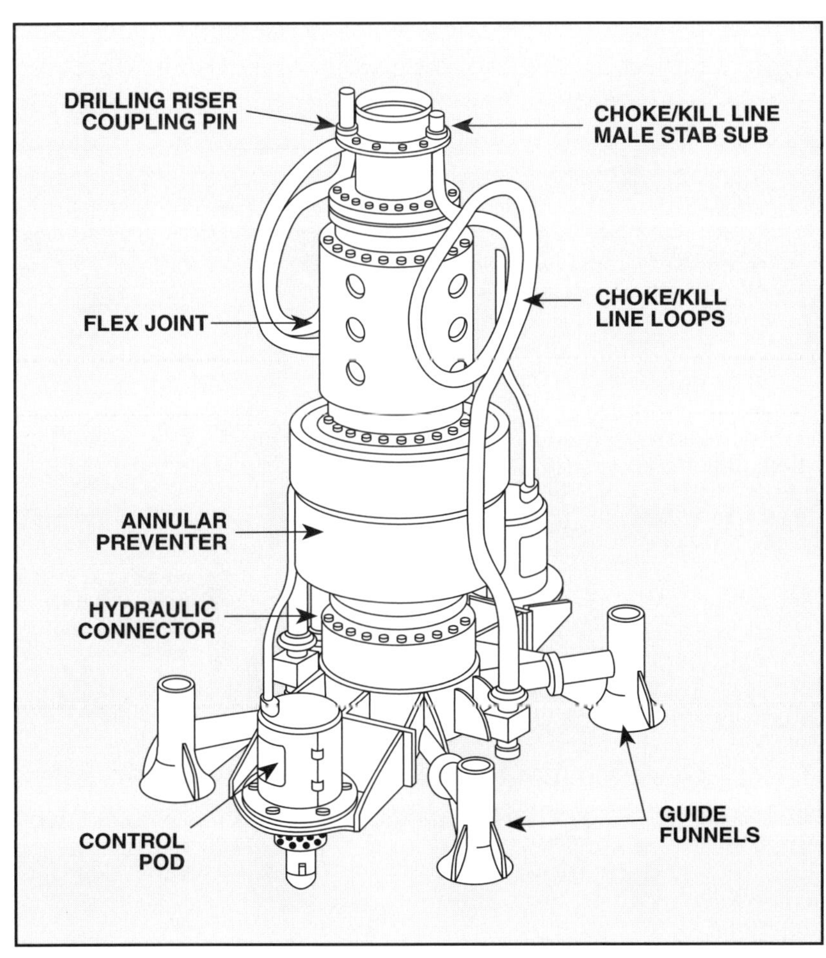

Figure 18. Lower riser marine package (LMRP) (Courtesy of ABB Vetco Gray)

Hydraulic Connector

A hydraulic connector allows the LMRP to be disconnected from the BOP stack. It connects to the upper mandrel of the BOP stack. The pressure rating of the hydraulic connector is equal to that of the annular preventer above it. (The operation of the connector is covered later.)

Flex and Ball Joints

A flex joint (fig. 19) or a ball joint is mounted between the annular preventer and the riser adapter on the LMRP. Both flex and ball joints flex, or bend laterally, to prevent excessive *bending moments* from being exerted on the marine riser and the LMRP and BOP components. (A bending moment is a force lateral movement creates on an object.) A flex joint typically allows 10 degrees of offset from vertical. The riser adapter can be connected to the top of the flex joint's neck by a flange, a hub, or by welding.

The hydrostatic pressure of the seawater outside the flex joint, as well as the hydrostatic pressure of the mud column inside the riser, subjects the joint to compression and tension. Moreover, overtensioning the riser from the surface produces tension loads.

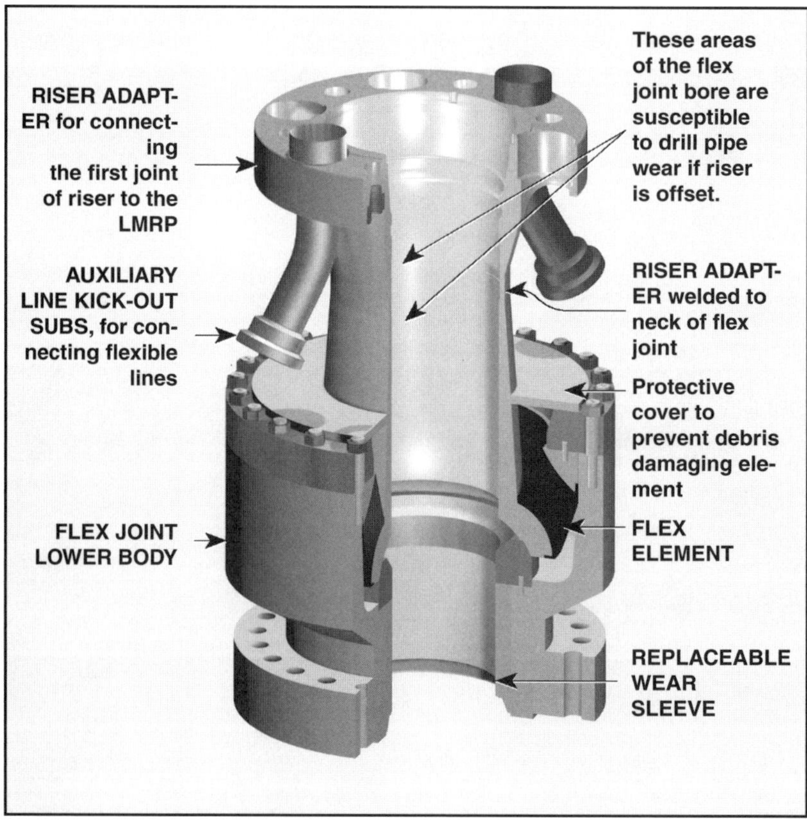

Figure 19. Flexible joint (Courtesy of Cameron)

Conversely, undertensioning the riser from the surface produces compression loads.

The rotational stiffness of flex joints makes them more effective than ball joints in controlling riser angles. Typically, the rotational stiffness of a flex joint is a nonlinear function of angle and it ranges from 10,000 to 30,000 foot pounds (ft•lb) or 13,558 to 40,672 newton metres (N•m) per degree of rotation. Rotational stiffness may also vary with temperature.

The flex element is manufactured by bonding laminations of synthetic rubber between stacks of spherical steel rings. The synthetic rubber provides the flexure and pressure sealing, and the steel plates provide the strength. Some designs provide a landing shoulder for a wear bushing that protects the body against damage caused by the drill pipe rubbing against the wall. Damage from drill pipe rubbing against the wall is called key seating and, if excessive, can lead to failure.

One type of wear bushing, or sleeve, can be removed or installed with a special handling tool that is run on drill pipe. Other types of wear bushing can only be replaced during periodic overhaul of the flex joint.

A ball joint (fig. 20) is a forged-steel ball and socket containing a cylindrical neck extension with a riser adapter attached at the top of the neck. The ball and socket has several seals that keep drilling mud out of the ball and socket. They also seal off the control fluid that provides pressure balance.

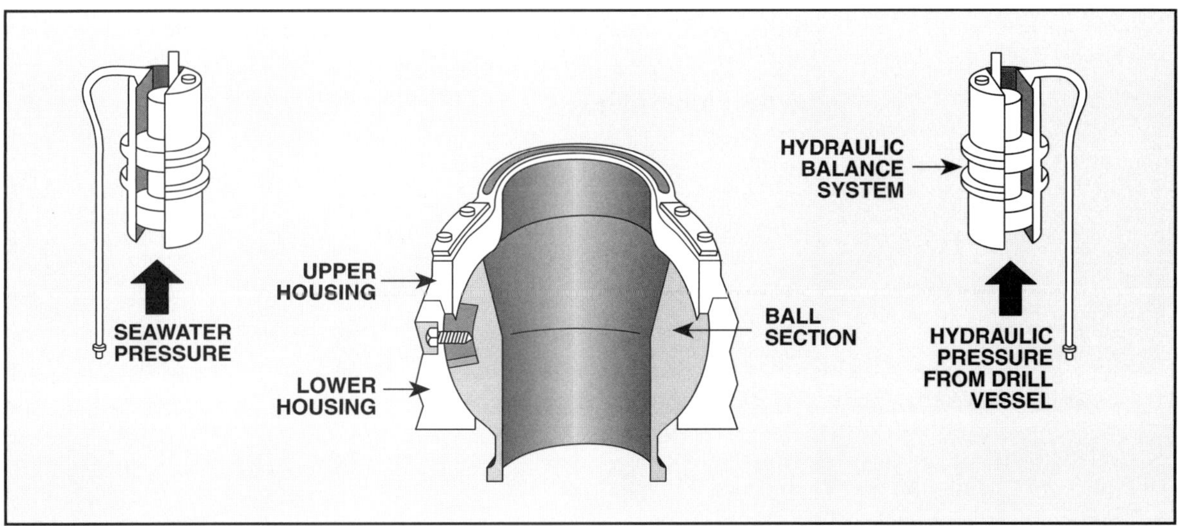

Figure 20. Ball joint (Courtesy of ABB Vetco Gray)

The same tension and compression loads that affect a flex joint also affect a ball-and-socket joint. So, to ensure freedom of movement without excessive torque, the ball has to be pressure balanced in relation to the socket. Without pressure balance, the ball would be pulled hard against the socket, and excessive torque would be required for the ball to move. Control fluid is applied between the ball and socket to keep the surfaces floating and allow free movement. However, it can be difficult to maintain the correct balance, and without the correct balance, the riser adapter can part because the ball joint cannot rotate freely. Thus, flex joints are preferred.

Flexible Piping

Flexible piping connects the choke and kill lines and the hydraulic conduit lines to the LMRP. Flexible piping allows the flex joint on the LMRP to move freely. Most rigs use Coflexip high-pressure hoses whose pressure rating is the same as the ram preventers. The hoses extend around the LMRP components from kick-out subs on the riser adapter to the LMRP baseplate (see fig. 19). The termination at the riser adapter is called a kick-out sub because it forms an angle of about 45 degrees to kick the hose out past the LMRP components.

At the baseplate, the hose is terminated to the top of a male stub that stabs into a female receptacle mounted on the BOP stack's upper receiver plate. The receptacle and upper receiver plate complete the seal between the LMRP and the BOP for the choke and kill lines. In some cases, the flexible lines terminate on the LMRP's baseplate to miniature hydraulic collet connectors (fig. 21). Coiled flexible pipe loops may be used as an alternative to the Coflexip hoses.

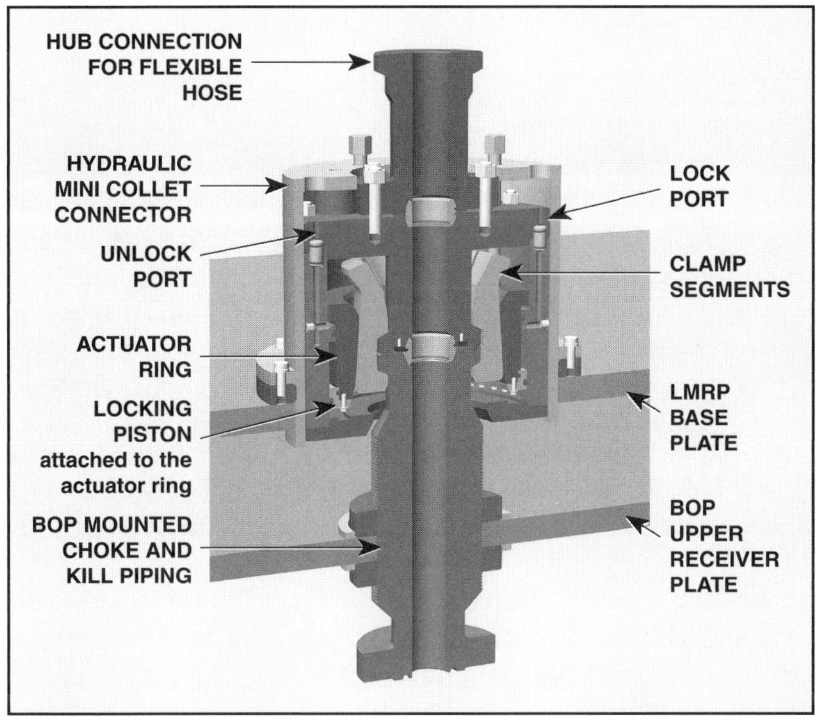

Figure 21. Miniature collet hydraulic connector (Courtesy of ABB Vetco Gray)

Riser Adapter

A riser adapter on top of the flex joint provides a transition point from the LMRP to the first riser joint. A riser adapter includes—

- a side entry for the mud booster line,
- a check valve or gate valve for closing off the mud booster line,
- a riser connection, and
- the kick-out subs on the choke and kill lines and on the hydraulic conduit line.

The riser adapter may be connected to the flex joint by welding directly to the neck, by a flange connection, or by a hub-and-clamp connection. A wear sleeve is often provided in the bore of the riser adapter.

Annular BOP

An annular preventer is also located on the LMRP. This manual covers annular BOPs in the blowout preventer section.

To summarize—

A marine riser system—

- provides a fluid conduit to and from the wellbore—that is, it extends the wellbore from the subsea BOP to the drilling rig;
- supports auxiliary lines, such as high-pressure choke and kill lines, mud booster lines, and hydraulic conduits;
- guides the drill stem and other tools from the drilling rig to the wellhead on the seabed; and
- provides a means of running and retrieving the BOP assembly from the surface to the wellhead on the seafloor.

Factors that affect marine riser design include—

- dynamic and axial loads while running and retrieving the riser system and BOP assembly;
- lateral forces from currents and vessel offset;
- cyclic forces from wave and vessel motion;
- vortex induced vibrations (VIVs);
- axial loads created by the weight of the riser system itself, the weight of the drilling fluid inside the riser, and the additional weight of freestanding pipe within the riser;
- axial tension from the tensioning system at the surface;
- dimensional requirements such as outside diameter and wall thickness of the main riser pipe;
- outside diameter (OD) and pressure rating of the choke and kill lines;
- OD and pressure rating of the mud booster line;
- OD and pressure rating of the hydraulic conduit line;
- method and makeup time of joint connections;
- storage and handling requirements; and
- lifetime operating and maintenance costs.

A typical marine riser system consists of—

- surface tensioning equipment;
- a diverter system;
- a telescopic, or slip, joint;
- individual riser joints;
- high-pressure choke and kill lines;
- hydraulic conduit lines;
- a mud booster line;
- a riser fill-up valve;

- a termination spool;
- an instrumented riser joint;
- a lower marine riser package (LMRP), which consists of flexible piping for the choke and kill, hydraulic conduit, and mud booster lines;
- a flex joint;
- one or two annular BOPs; and
- a hydraulic connector for connecting the LMRP to the BOP stack.
- A riser joint is a large-diameter, seamless or welded high-strength pipe that has connectors at each end. Special clamps hold the auxiliary lines to the pipe.
- Some common BOP IDs and riser joint ODs are—

BOP ID, in. (mm)	Riser OD, in. (mm)
13⅝ (346)	16 (406.4)
16¾ (425.5)	18⅝ (473.1)
18¾ (476.3)	21 (533.4)
21¼ (539.8)	24 (609.6)

- Manufacturers offer riser joints in several lengths. Typical lengths are 50, 65, 70, 75 and 90 ft (15.24, 19.81, 21.34, 22.86, and 27.43 m). Lengths of 70 ft (21.34 m) and longer are most often used in deepwater applications. For shallower water operations, 50-ft (15.24-m) lengths are common because they are easier to handle than longer joints.

Riser joints—

- are made from X80-grade steel with a yield strength of 80,000 psi or 551,600 kPa.
- have a wall thickness of ½, ⅝, 11⁄16, ¾, 1, and 1¼ in. (12.7, 15.9, 17.5, 19.1, 25.4, and 31.8 mm).
- should have an internal pressure rating that is at least equal to the working pressure of the surface diverter system, plus the maximum difference in hydrostatic pressure between the drilling fluid inside the riser and the seawater outside.
- usually have an automatic flooding valve in a deepwater riser system to prevent collapse of the riser when accidental evacuation of the mud in the riser occurs.

Riser connectors—

- are either the dog type or the bolted-flange type.
- of the bolted-flange type are designed for drilling in water depths of 7,000 ft (2,100 m) or more and have tensile load ratings of 3.5 million pounds (lb) or 1,557,500 decanewtons (dN).
- of the dog type have a tensile load rating of 2 million lb (890,000 dN).

Termination spools—

- are used to overcome height restrictions.

Telescopic joints—

- are installed at the top of the marine riser system.
- compensate for vertical movement (heave) of the vessel.
- provide a means of connecting the diverter assembly to the riser.
- provide terminations for the riser auxiliary lines to flexible hoses at the drilling vessel.
- provide attachment points for the riser tensioning system.
- are made up of an inner and an outer barrel. The outer barrel is attached to the top joint of the marine riser assembly. The inner barrel is attached to the diverter assembly's flex joint. Riser tensioning lines are attached either to a swivel tensioner ring or to a hydraulic tensioner ring on the outer barrel. A swivel tensioner ring is bolted to the outer barrel and can rotate around the outer barrel.
- have a stroke length from 45 to 65 ft (about 15 to 20 m) depending on the rig's requirements. A 45-ft (15-m) stroke is normally sufficient for shallow water, where the environment is relatively mild. A 65-ft (20-m) stroke is used for deepwater operations.

Auxiliary lines—

- include choke and kill lines, mud booster lines, and hydraulic conduit lines.
- should be placed asymmetrically around the riser flange.
- should meet the requirements of the National Association of Corrosion Engineers (NACE) Specification MR-01-75, which covers steels intended for hydrogen sulfide (H_2S) service.

- should have line couplings able to seal against full pressure, while allowing for relative motion between the box and pin.
- have shoulders on the box and pins of the choke and kill lines, as well as on the support flanges, to share a portion of the main riser pipe's load.

Choke and kill lines—

- extend the wellbore to the drilling rig when the subsea BOP is closed on a well kick.
- are made of X52, X65, or X80 steel, which have a yield of 52,000, 65,000, and 80,000 psi (358,540, 448,175, and 551,600 kPa) respectively.
- have an ID of 3 in. (76.2 mm) or 4½ in. (114.3 mm).

Booster lines—

- increase the return velocity of the fluid column in the riser.
- are connected to a pump on the rig that enables the driller to circulate drilling fluid from the surface to the bottom of the riser.
- are usually made of X52 or X65 steel, which have a yield of 52,000 and 65,000 psi (358,540 and 448,175 kPa) respectively.
- generally have an ID of 3 in. (76.2 mm).

Hydraulic conduit lines—

- supply hydraulic fluid to the BOP control system in deepwater operations.
- normally have a 2- or 3-in. (50.8- or 76.2-mm) ID.
- are normally manufactured from stainless steel to prevent rust from occurring and contaminating the fluid.

Handling tools—

- raise and lower the riser and telescopic joints.
- have the same tensile load rating as the riser system and can support the full load of the riser and BOP assembly.
- are either mechanical or hydraulic; a mechanical riser handling tool duplicates a riser joint connection. It is made up to a riser joint just as though a riser joint was being connected. A hydraulic riser handling tool reduces the time involved in making the connections. Hydraulic pressure forces a sleeve downward behind a split lock ring, which expands into a recess in the body of the riser connection.

Spiders—

- support the riser and BOP assembly as the crew picks up and adds each new joint to the string.
- consist of a base plate and top plate with a series of support dogs, or arms, between them. The dogs support the riser joint under the riser joint's support flange.
- may have a support gimbal between the rotary table and the riser spider in deepwater operations.

Buoyant riser modules—

- reduce the weight of the riser string.
- are made from syntactic foam.
- are attached to the outside of each riser joint.
- enable drilling in water depths of from 2,000 to 10,000 ft (600 to 2,000 m).

The lower marine riser package (LMRP) is made of—

- a riser adapter,
- a flex or ball joint,
- an annular BOP,
- BOP control pods,
- a hydraulic connector,
- flexible choke and kill lines, and
- choke and kill stabs.

The LMRP—

- is an interface between the riser system and the BOP stack that crew members can disconnect should the need arise.

Diverters

A diverter system (fig. 22) protects personnel and equipment by diverting the flow from shallow gas kicks overboard. Such gas flows can occur before a BOP stack can be installed because no casing is set in the well. A diverter does not shut in or halt well flow; rather, it diverts flow away from the rig. During normal drilling operations, the diverter directs the flow of mud returning from the marine riser into the rig's return flow line.

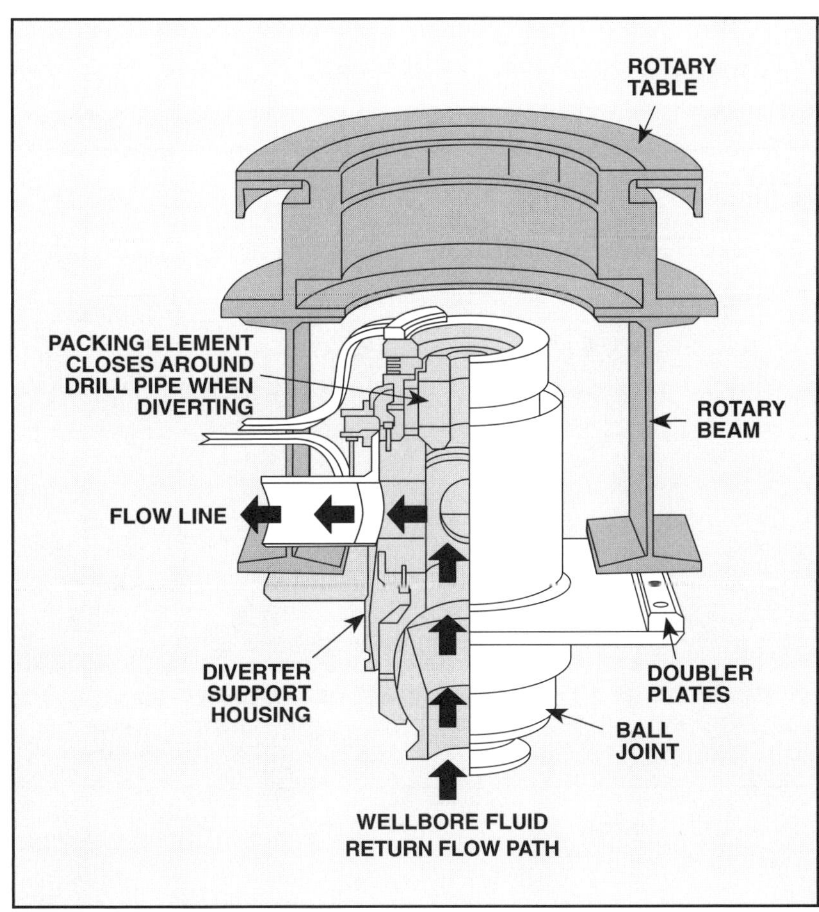

Figure 22. Diverter (Courtesy of ABB Vetco Gray)

A packing element in the diverter closes and seals around the kelly or drill pipe to prevent shallow gas kicks from venting on the drill floor. When the diverter element is closed, valves in the flow line and vent lines direct the flow directly overboard. The valves in the system are selectively sequenced through the diverter control system and operate in conjunction with the packing element. The selective sequencing prevents the flow's being shut in by operating each valve and the packing element in a predetermined sequence.

Shutting in a shallow gas kick creates back-pressure on the well that could break down the formation and cause a blowout around the conductor pipe. This condition can be extremely hazardous for a floating rig, as the escaping gas can cause the vessel to lose buoyancy and sink. Figure 22 shows an ABB Vetco Gray KFDS diverter system, which is typical of such systems. It cannot close on open hole and has to be selectively sequenced through the control system.

In some designs, such as the Vetco CSO (fig. 23), the diverter element is similar to that in an annular preventer. The spherical element in such a diverter can close on open hole and seal against full working pressure.

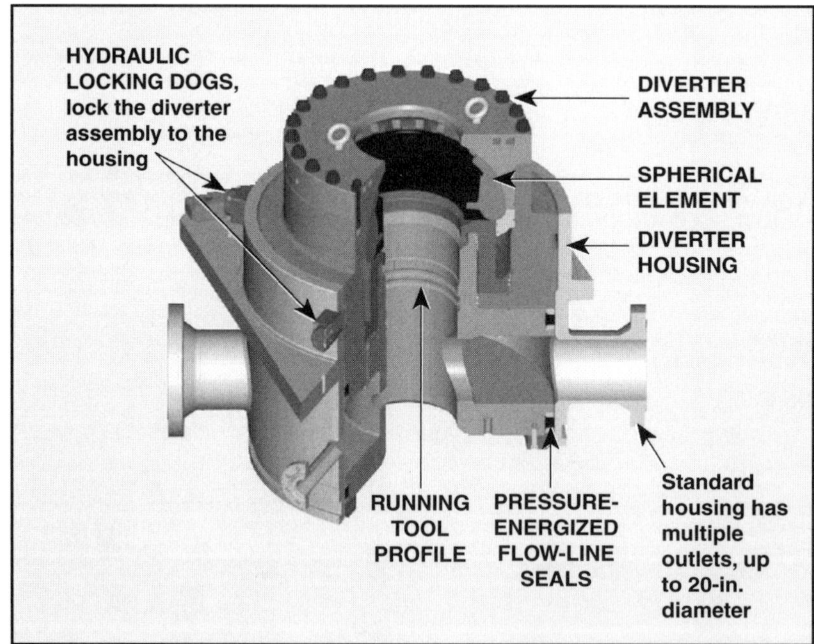

Figure 23. Diverter that can close on open hole as well as the drill stem (Courtesy of ABB Vetco Gray)

A diverter system generally consists of a permanently fixed housing, which is bolted to large support beams below the rotary table (see fig. 22). The mud flow line and overboard vent lines are permanently attached to the side outlets of the fixed diverter housing. The outlets of the overboard vent lines are normally positioned at opposite extremities of the vessel. Selecting the correct vent line through the valve system is based solely on wind direction.

The diverter assembly, which contains a packing element, flow-line seals, and lockdown mechanisms (see fig. 23), is run and retrieved with the riser system and has its own handling tool. When installed in the fixed housing, the assembly is locked in place by hydraulically-operated locking dogs. Pressure-energized flow-line seals seal the annulus between the assembly and the fixed housing. A flex joint on the assembly's lower section connects to the telescopic joint's inner barrel and prevents excessive bending forces from being exerted on the fixed housing and the telescopic joint's inner barrel.

When only conductor casing (or some other type of casing) is set to a shallow depth, the BOP stack usually is not connected to the conductor string's casinghead. As mentioned earlier, shutting in a well with only conductor casing set will likely fracture the formation below the casing shoe. So, crew members connect the marine riser to the casinghead using a special pin connector and flex joint assembly (fig. 24). They then install the diverter system on top of the marine riser.

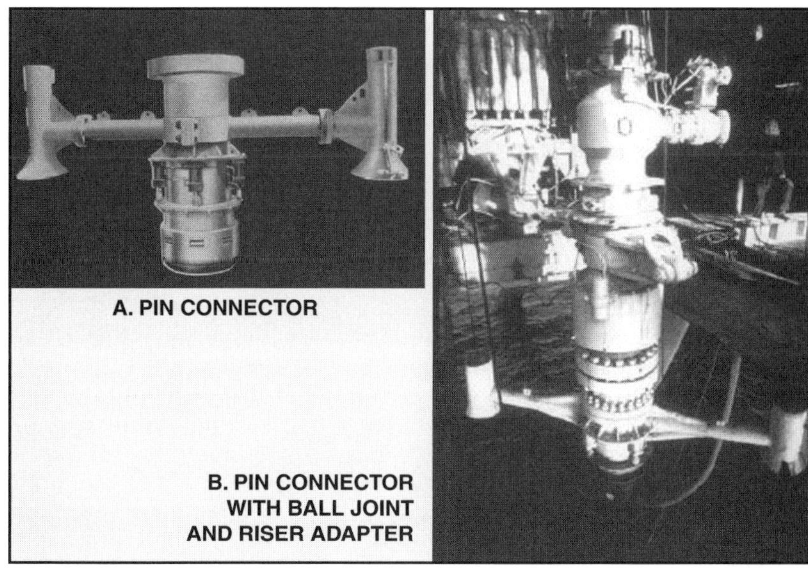

Figure 24. Special pin connect (A) and connector with ball joint and riser adapter (B) (Courtesy of ABB Vetco Gray)

To summarize—

Diverters—

- protect personnel and equipment by diverting the flow from shallow gas kicks overboard.
- do not shut in or halt well flow; rather, a diverter diverts flow away from the rig.
- have a diverter element, which, when closed, allows valves in the flow line and vent lines to direct the flow overboard.

Riser Tensioning Systems

The weight of the marine riser system must be supported while it is deployed. Otherwise, the weight of the joints at the top of the string can crush the joints below. If not supported, a conventional marine riser buckles in water depths greater than 200 to 300 ft (60 to 90 m).

A riser tensioning system (fig. 25) must support the weight of the longest riser string at the rig's deepest operating water depth.

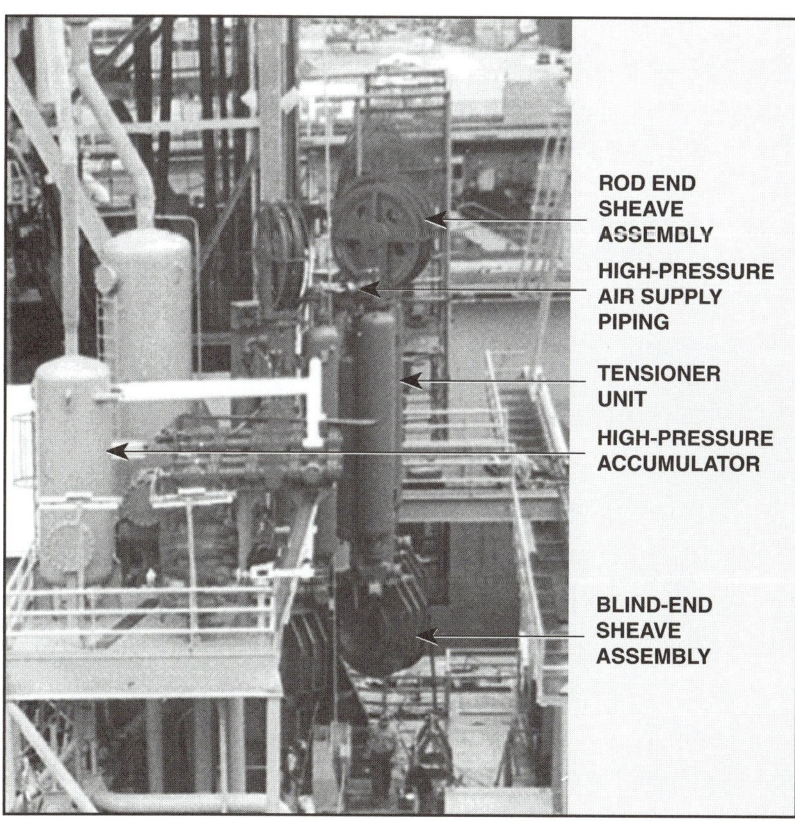

Figure 25. Riser tensioning system (Courtesy of Shaffer)

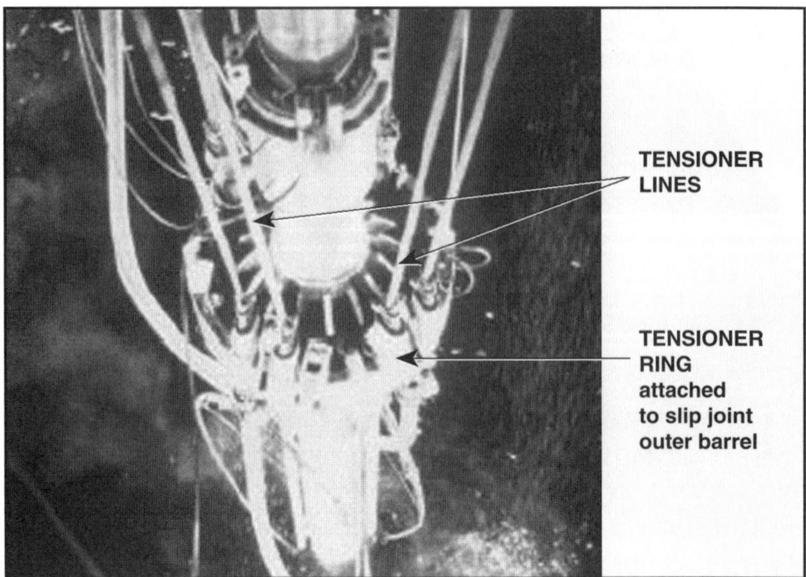

Figure 26. Tensioner lines connected to a riser tensioner ring (Courtesy of Shaffer)

The tensioning system supports weight by applying tension to the outer barrel of the telescopic joint. Tensioner lines are connected to a tensioner ring or to fixed padeyes on the outer barrel (fig. 26). The applied tension must remain constant while compensating for vessel heave.

Many variables determine the amount of tension required, including such items as water depth, mud weight inside the riser, wave and current forces, vessel motion, and buoyancy effects. Personnel can use computer modeling to determine the correct amount of tension. Major components of a hydropneumatic riser-tensioning system include—

- hydropneumatic cylinders and sheave assemblies,
- accumulators and system air-pressure vessels,
- control panel and piping manifold,
- high-pressure air compressors, and
- standby air-pressure vessels.

Tensioner units are usually located on the drill floor's substructure (see fig. 25). Individual units consist of a hydro-pneumatic cylinder assembly. A high-pressure compressor, backed up by a standby air system, supplies air to the hydropneumatic cylinders. Installers reeve wire rope through dual sheaves, which are mounted on the piston head and the cylinder's blind end.

One end of the wire rope is anchored at the cylinder's blind end, and the other end is attached to a padeye on the telescopic joint's outer barrel. The wire rope features four-part reeving—that is, effective wireline travel is four times the piston travel. So, a 12½-ft (4-m) stroke of the piston rod equals 50 ft (16 m) of line travel. The high-pressure air acting on the piston generates force, which is converted to tension on the wire rope through the sheaves.

Because a constant tension must be maintained on the riser and because air pressure creates that tension, air pressure must also be kept constant. To keep the air pressure constant, tensioner units are connected to large-volume air-pressure vessels, which maintain the air pressure to within 3 percent of the selected pressure over the full length of the stroke. The rod end of the cylinder above the piston is filled with fluid under a pressure of 20 to 40 psi (140 to 280 kPa). This fluid pressure lubricates the cylinder, dampens piston motion, and provides rod speed control in the event of a broken wire rope.

The dynamic tension limit is the maximum allowable pressure multiplied by the effective hydraulic area, divided by the number of line parts. The equation is—

$$D_{tl} = P_a \times A_{cyl} \div N_{lp} \qquad \text{(Eq. 3)}$$

where

D_{tl} = dynamic tension limit

P_a = maximum allowable system operating pressure

A_{cyl} = effective hydraulic area

N_{lp} = number of line parts.

When installing a riser tensioning system, personnel should keep several points in mind.

- The number of tensioner units required for a specific rig and riser depends on the riser's size and length (its total weight), the mud weight, the amount of suspended pipe within the riser, ocean currents, wave heights and periods, vessel offset from above the well, and vessel motion.
- All components in a riser tensioning system, including the piping, should be designed for the system's maximum working pressure.
- The tensioner system should be designed to allow at least two of the units to be out of service for maintenance or repair.

- The maximum tension setting should not be more than 80 percent of the dynamic tensioning limit (D_{tl}). Maintaining maximum tension at no more than 80 percent of D_{tl} ensures that the maximum tension, including dynamic variations, will be less than D_{tl}.
- Tensioner systems, depending on the manufacturer, are rated for 2,400 or 3,500 psi (16,800 or 24,500 kPa) air pressure. They can be rated from 80,000 to 200,000 lb (35,600 to 89,000 dN) of single-line tension. The wire rope used in tensioner systems has ODs ranging from 1¾ to 2½ in. (44.5 to 63.5 mm).
- Because the wire rope in riser tensioner systems undergoes rough service, rig supervisors should implement a slip-and-cut program, similar to that used with drilling line, to avoid wire rope breakage.

To summarize—

Riser tensioning systems—

- support the weight of the longest riser string at the rig's deepest operating water depth.
- support the weight by applying tension to the outer barrel of the telescopic joint.
- consist of hydro-pneumatic cylinders and sheave assemblies, accumulators and system air-pressure vessels, control panel and piping manifold, high-pressure air compressors, and standby air-pressure vessels.
- require that the following points to be kept in mind:
  1. The number of tensioner units required for a specific rig and riser depends on the riser's size and length (its total weight), the mud weight, the amount of suspended pipe within the riser, ocean currents, wave heights and periods, vessel offset from above the well, and vessel motion.
  2. All components in a riser tensioning system, including the piping, should be designed for the system's maximum working pressure.
  3. The tensioner system should be designed to allow at least two of the units to be out of service for maintenance or repair.

---

4. The maximum tension setting should not be more than 80 percent of the dynamic tensioning limit (D_{tl}). Maintaining maximum tension at no more than 80 percent of D_{tl} ensures that the maximum tension, including dynamic variations, will be less than D_{tl}.
5. Tensioner systems, depending on the manufacturer, are rated for 2,400 or 3,500 psi (16,800 or 24,500 kPa) air pressure. They can be rated from 80,000 to 200,000 lb (35,600 to 89,000 dN) of single-line tension. The wire rope used in tensioner systems has ODs ranging from 1¾ to 2½ in. (44.5 to 63.5 mm).
6. Because the wire rope in riser tensioner systems undergoes rough service, rig supervisors should implement a slip-and-cut program, similar to that used with drilling line, to avoid wire rope breakage.

Guideline Tensioners

In water depths no greater than about 5,000 ft (1,500 m), guidelines may be used to guide the BOP and LMRP assembly to the wellhead on the seabed. Four wire-rope guidelines form the corners of a square that extend from the vessel to guideposts on a guide base at the seabed. Each wire rope is attached to a guideline tensioner through double sheaves on the piston rod and cylinder's blind end. The guideposts on the guide base are 6 ft (2 m) from the center of the wellbore, and form a square whose sides are about 8½ ft (2.5 m) long. Guidelines are inserted into funnels at each corner of the BOP and LMRP frames.

When the BOP and LMRP are run, the guidelines are tensioned sufficiently to (1) prevent rotation of the BOP and LMRP assembly, (2) overcome the lateral loads imposed on the guidelines by vessel movement and current forces, and (3) provide alignment with the guideposts on the guide base.

Once the BOP assembly has been lowered over the guideposts and landed on the wellhead, the guideline tension can be reduced to a point sufficient just to support the weight of the wire rope.

If the LMRP has to be disconnected from the BOP, the guideline tensioners are used to guide the LMRP back over the BOP guideposts.

The guideline tensioners may also be used to guide other equipment to and from the seabed. For example, they may guide TV cameras to monitor riser angle or to see whether hydrates are forming. Guideline tensioners may also be used to guide control pods to and from the LMRP. Experience has shown that even the heaviest loads can be guided with guideline tension at the surface set at 6,000 to 12,000 lb (2,670 to 5,340 dN).

Equipment in the guideline tensioning system is basically the same as that found in the marine riser tensioning system. The equipment must not only provide the required tension to the guidelines, but also compensate for vessel motion.

Guideline tensioning equipment consists of a hydropneumatic cylinder assembly that compressors and a standby air pressure vessel supply with high-pressure air. The wire rope is reeved through dual sheaves mounted on the piston head and the cylinder's blind end. One end of the wire rope is stored and anchored on a retrieving winch at the cylinder's blind end; the other end is attached to the guide base. Because the wire rope is reeved four times through the dual sheaves, effective wireline travel is four times the piston travel. Four-part reeving allows 40-ft (12-m) line travel from a 10-ft (3-m) stroke piston rod. The high-pressure air on the piston generates a force, which the sheaves convert to tension on the wire rope.

Since constant tension must be maintained on the guideline and since that tension is created by air pressure, air pressure must also be kept constant. To keep the air pressure constant, the guideline tensioner units are connected to large-volume air pressure vessels, which maintain the air pressure to within 3 percent of the selected pressure over the full stroke length.

Guideline tensioners are rated from 16,000 to 20,000 lb (7,120 to 8,900 dN) of tension and require an air supply pressure of 2,400- to 3,500-psi (16,500- to 24,500-kPa).

Fluid, under a pressure of 20 to 40 psi (140 to 280 kPa), fills the rod end of the cylinder above the piston. This pressurized fluid lubricates the cylinder, dampens piston motion, and controls rod speed in the event of a broken wire rope.

If the guideline tensioner's wire rope breaks, it usually occurs at the guideposts on the subsea guide base. Insufficient tension as the BOP is lowered over the guideposts can cause the guidelines to break. Also, if the BOP moves laterally in relation to the guideposts as it is being lowered, the BOP frame can cut the wire rope. The BOP funnels have smooth surfaces and curvatures to match the radius of the wire-rope sheaves, which are appropriate for the size of the wire rope in use. Guideposts on BOP stacks and guide bases have smooth internal surfaces and curvatures to minimize guideline damage. Guideline wire ropes normally require replacement because of cuts and kinks rather than wear or fatigue.

To summarize—

Guideline tensioners—

- put tension on guidelines, which are used on floating rigs drilling in waters no deeper than about 5,000 ft (1,500 m).
- put tension on the guidelines, which consist of four wire ropes that extend from the vessel to guideposts on a guide base at the seabed.
- are tensioned sufficiently to (1) prevent rotation of the BOP and LMRP assembly, (2) overcome the lateral loads imposed on the guidelines by vessel movement and current forces, and (3) provide alignment with the guideposts on the guide base.
- are also used to guide other equipment to and from the seabed. For example, they may guide TV cameras to monitor riser angle or to see whether hydrates are forming.
- consist of a hydropneumatic cylinder assembly that compressors and a standby air pressure vessel supply with high-pressure air, and a wire rope that is reeved through dual sheaves mounted on the piston head and the cylinder's blind end. One end of the wire rope is stored and anchored on a retrieving winch at the cylinder's blind end; the other end is attached to the guide base.
- are rated from 16,000 to 20,000 lb (7,120 to 8,900 dN) of tension and require an air supply pressure of 2,400- or 3,500-psi (16,500- to 24,500-kPa).

Subsea Wellheads and Casing

The subsea wellhead and casing are a system—a system that forms the foundation for drilling a successful well. This system also links the well to the BOP equipment. A structural casing string is set and cemented in place to support the BOP assembly at the seabed. This structural casing also supports subsequent casing strings that are run in the well.

As the well is drilled, crew members run casing that is smaller in diameter and has higher pressure ratings than previous casing strings. Table 2 gives a typical casing program.

Table 2
Casing Strings Run into a Well From Start to Finish with Typical Hole and Casing Sizes

Casing String	Hole Size, in. (mm)	Casing Size, in. (mm)
Structural	36 (914.4)	30 (762.0)
Conductor	26 (660.4)	20 (508.0)
Surface	13⅜ (339.7)	17½ (444.5)
Intermediate	12¼ (311.2)	9⅝ (244.5)
Production	8½ (215.9)	7 (177.8)

Prior to drilling the 36-in (914.4-mm) hole, the operator surveys the seabed to determine its condition and to look for obstructions. If necessary, the rig crew runs a temporary guide base with four guidelines attached to it. Figure 27 shows a temporary guide base about to be lowered into the water. Note the guidelines and the sacks of barite on the base to provide weight. A special running tool for the guide base is made up on drill pipe.

Figure 27. Temporary guide base (Courtesy of ABB Vetco Gray)

Once the temporary guide base has been run and landed on the seafloor, and the running tool retrieved, the 36-in. (914.4-mm) drilling assembly is lowered to the seabed using the guidelines. Then, crew members drill the hole through the center of the temporary guide base. They lower the 30-in. (762.0-mm) structural casing into the hole and cement it in place. Prior to cementing the structural casing, personnel check to ensure it is vertical to within 1 degree. If it is at an angle of more than 1 degree, the situation should be corrected because if the BOP assembly is landed at an angle greater than 1 degree, the drill string can rub against the BOP components and damage them.

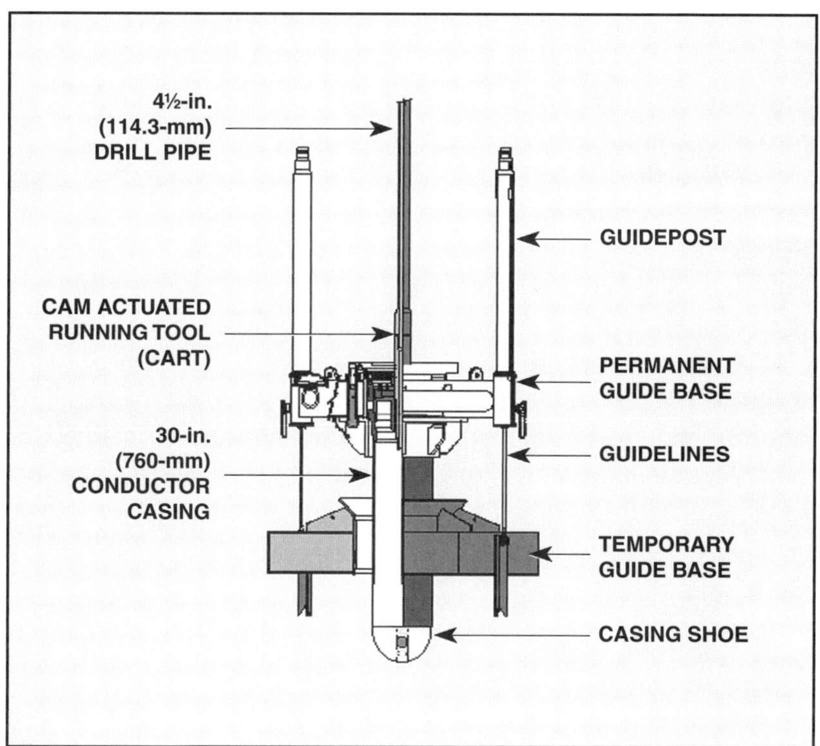

Figure 28. Permanent guide base being landed on the temporary guide base (Courtesy of ABB Vetco Gray)

The last joint of 30-in. (762.0-mm) casing to be run has a specially profiled, 30-in. (762.0-mm) low-pressure housing welded to it. Figure 28 shows the 30-in. (762.0-mm) casing and permanent guide base with four guideposts being lowered on drill pipe through the temporary guide base. The cam-actuated running tool (CART) allows the drill pipe to be disconnected from the permanent guide base after it lands in the temporary guide base. After the crew lands the permanent guide base on the temporary guide base, they cement the casing in place.

Figure 29 shows the 30-in. (762.0-mm) conductor housing and CART. The OD of the housing is profiled to accept the permanent guide base, which, along with the guidelines, guides the BOP to the wellhead. The housing's ID accepts the CART that enables the crew to land and latch the housing. It also accepts a low-pressure pin connector that can be run with a flex joint and marine riser to allow returns to the surface while drilling the 26-in. (660.4-mm) section of hole. The bottom of the housing's ID has a landing shoulder into which an 18¾-in. (476.3-mm) high-pressure wellhead fits.

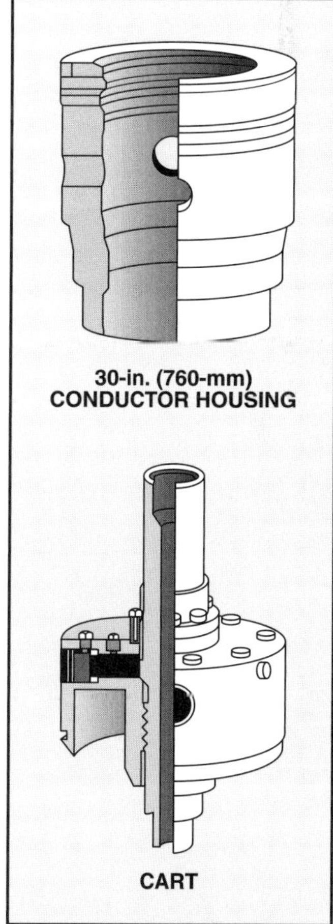

Figure 29. Conductor housing and CART

It is vitally important to the entire drilling operation that the structural casing be securely cemented in place, because the weight of all subsequent casing strings and the BOP are transmitted to the formation through the 30-in. (762.0-mm) housing.

Once the 30-in. (762.0-mm) structural casing is set and cemented, crew members make up and run a 26-in. (660.4-mm) drilling assembly into the hole. They then drill the 26-in. (660.4-mm) hole and run and cement the 20-in. (508.0-mm) casing and the 18¾- in. (476.3-mm) high-pressure wellhead.

Crew members weld the 18¾-in. (476.3-mm) high-pressure wellhead (fig. 30) to the last joint of 20-in. (508.0-mm) casing. This housing has a heavy wall with strengthening longitudinal ribs to resist bending moments transmitted through the marine riser, flex joint, and BOP assembly. A high-pressure wellhead—

- provides the foundation for the BOP stack,
- supports and houses subsequent casing strings, and
- provides a wellbore seal-and-locking arrangement between the surface casing and BOP stack.

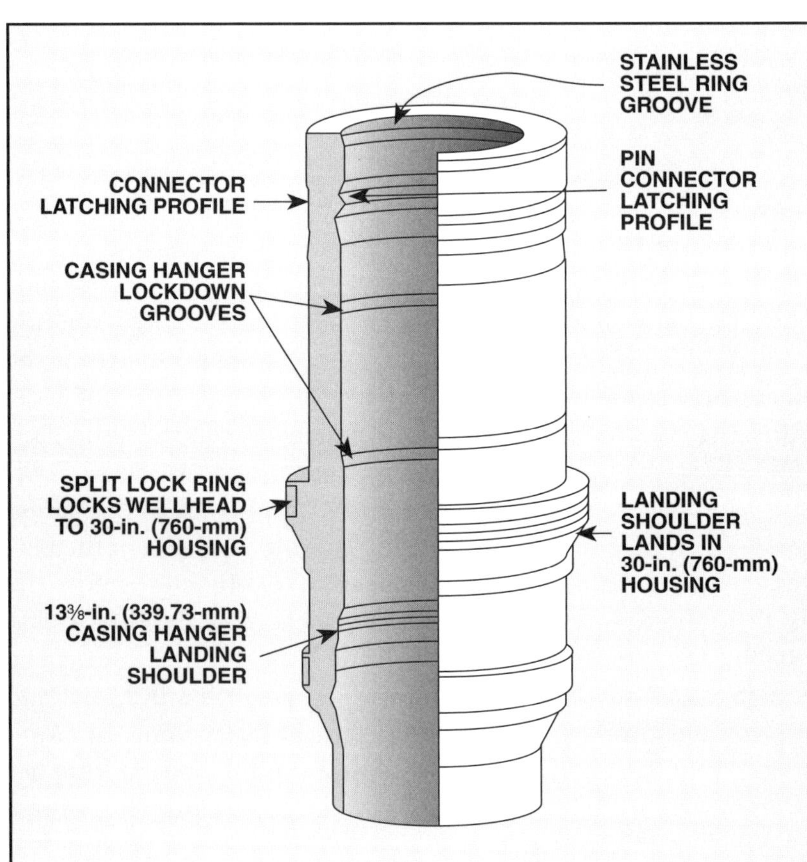

Figure 30. An 18¾-inch (476.3-mm) high-pressure wellhead (Courtesy of ABB Vetco Gray)

A landing shoulder on the 18¾-in. (476.3-mm) wellhead's OD fits inside a support shoulder inside the 30-in. (762.0-mm) housing. A spring-loaded latch mechanism snaps into a locking groove on the 30-in. (762.0-mm) housing to lock the wellhead in place. On the OD near the top of the 18¾-in. (476.3-mm) wellhead is a 45-degree connector-latching profile that the BOP connector locks to. On the ID near the top, the wellhead is profiled to accept a running tool, a stainless steel ring groove, a casing-hanger support shoulder, and various seals.

After the 20-in. (508.0-mm) casing and 18¾-in. (476.3-mm) high-pressure wellhead are run and cemented, crew members run the BOP assembly on marine riser pipe and land the BOP on the 18¾-in. (476.3-mm) wellhead. Once they confirm that the BOP is properly landed on the wellhead, the BOP connector is hydraulically locked. The crew can use a remotely operated vehicle (ROV) with a camera to check for full locking of the connector by observing the indicator rod travel. An overpull of approximately 50,000 lb (22,250 dN) above the landing weight is then applied to confirm lock down. Figure 31 shows the connector landed and locked to the wellhead.

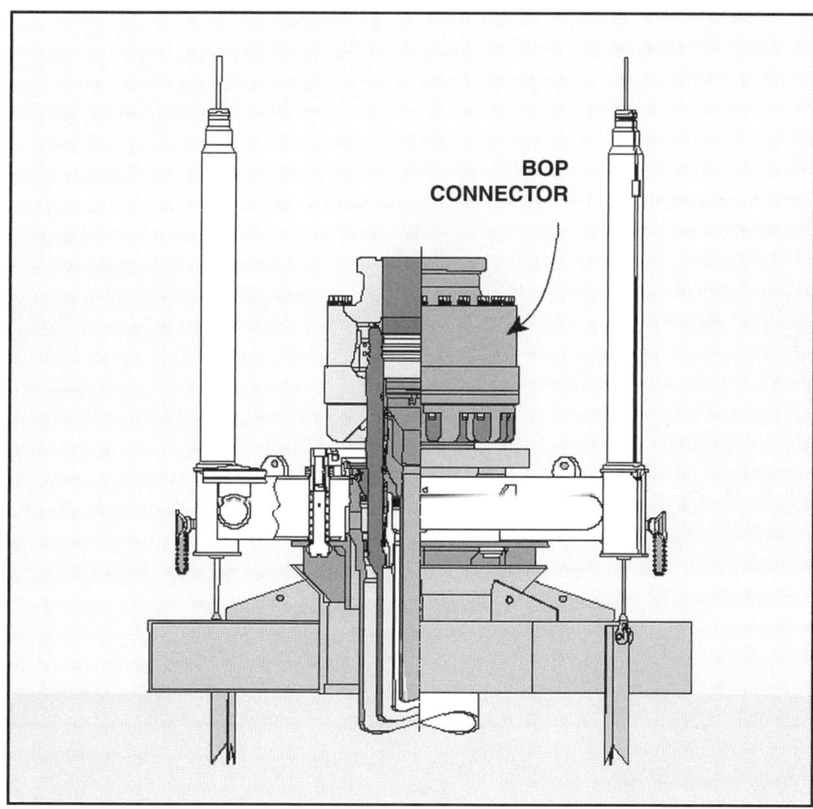

Figure 31. Connector locked and landed on subsea wellhead (Courtesy of ABB Vetco Gray)

Once the BOP has been run and locked to the wellhead, the crew runs a test plug (fig. 32) into the wellhead. The test plug's shoulder firmly seats in the wellhead and the O-rings form a seal to isolate the low-pressure casing. With the low-pressure casing protected from high pressures, crew members carry out a pressure test of the BOP components to confirm the wellhead-to-connector gasket is properly sealing.

After the crew runs and tests the BOP, they make up and lower into the well a 17½-in. (444.5-mm) drilling assembly to drill the next section. After drilling the 17½-in. (444.5-mm) section, crew members run 13⅜-in. (339.7-mm) casing with full well-control protection from the BOP assembly should a well kick occur. They land the 13⅜-in. (339.7-mm) casing hanger on the bottom shoulder of the 18¾-in. (476.3-mm) high-pressure wellhead. All loads are then transmitted to the 30-in. (762.0-mm) housing.

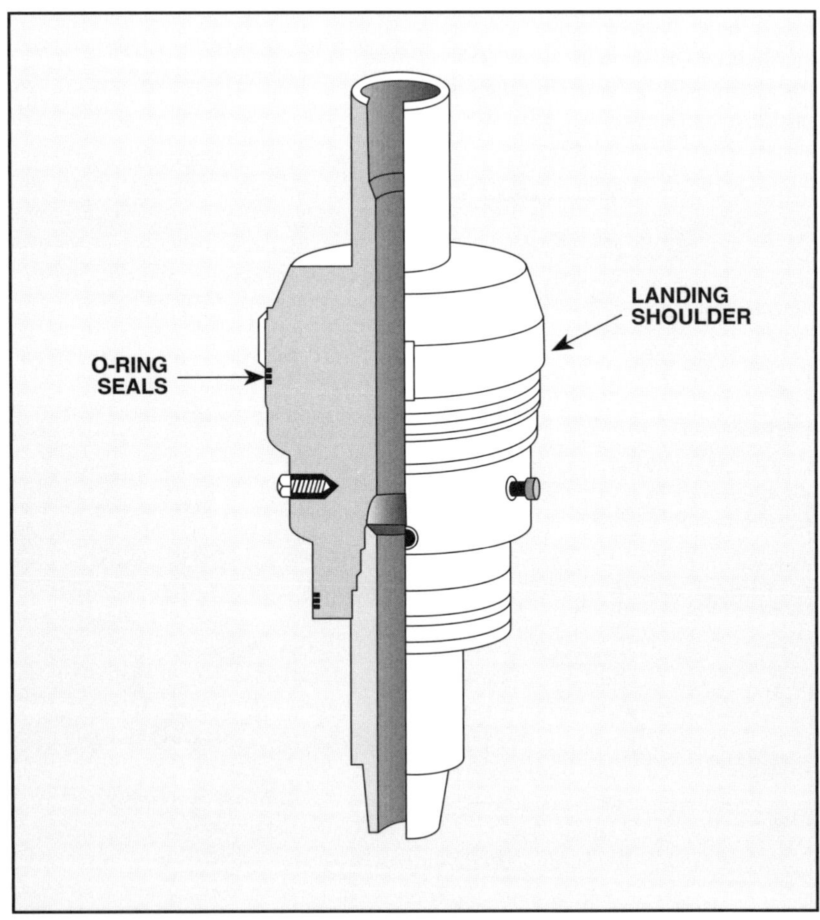

Figure 32. BOP test tool (Courtesy of ABB Vetco Gray)

After landing the 13⅜-in. (339.7-mm) casing hanger, crew members drill subsequent holes and run and land the casing strings in the 18¾-in. (476.3-mm) wellhead. Each casing string has a special hanger, which lands on the previous hanger. For example, the 9⅝-in. (244.5-mm) casing hanger lands in the 13⅜-in. (339.7-mm) hanger and so on.

Each casing hanger (fig. 33) has a metal-to-metal assembly to seal the annulus between each casing string and to isolate the previous casing from the higher wellbore pressures encountered at greater depths. The seal assemblies are tested for pressure integrity when they are initially installed. The running tool not only allows crew members to run and land the casing, but also serves as a test tool. It seals the bore of the casing hanger to isolate the casing string from the test pressure.

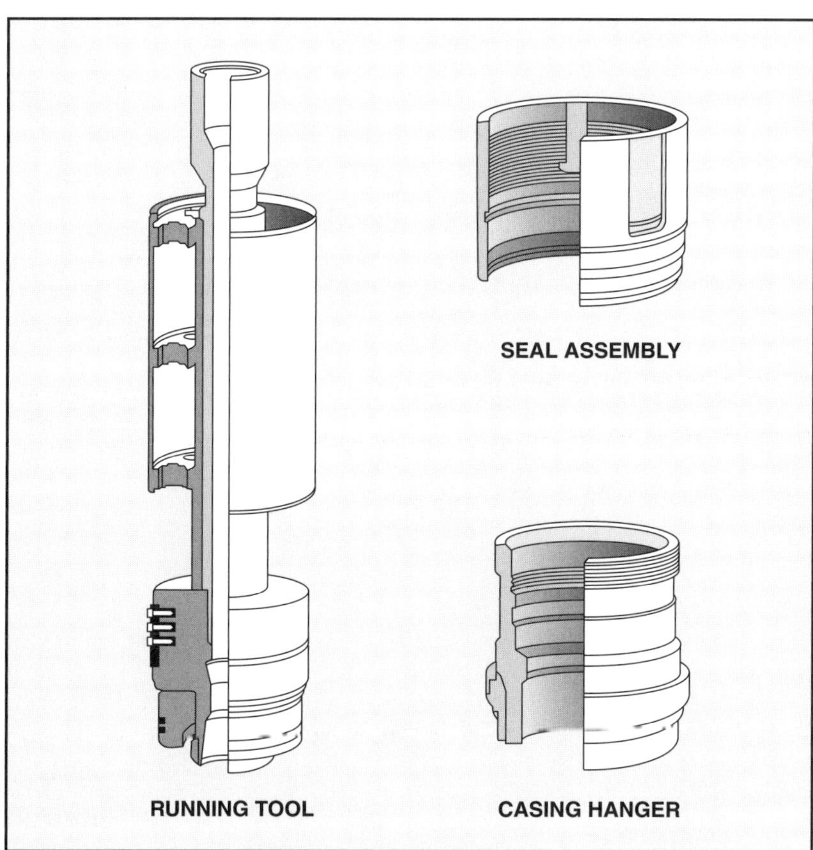

Figure 33. Casing hanger, seal assembly, and running tool (Courtesy of ABB Vetco Gray)

To test for pressure integrity, the driller closes the pipe rams and pumps a test fluid through the choke or kill line. It is important to closely monitor the amount of fluid pumped. Should the seal assembly fail, the test fluid has a flow path into the previous casing string, which has a lower pressure rating. If full test pressure is applied, and fluid flows into a previous casing string, this casing string could be damaged. The casing hangers also have a lockdown ring that allows them to be locked in place if required.

As each new section of the hole is drilled, the exposed sealing areas of the wellhead are protected by removable wear bushings. Crew members install and remove these wear bushings with special handling tools they run on drill pipe. Figure 34 is a cutaway view of casing hangers of various sizes landed on the different wellheads. Note the 9⅝-in. (244.5-mm) wear bushing at top. The 13⅜-in. (339.7-mm) seal assembly is energized by rotation of the pack-off nut. Once energized, it seals the annulus between the two casing strings.

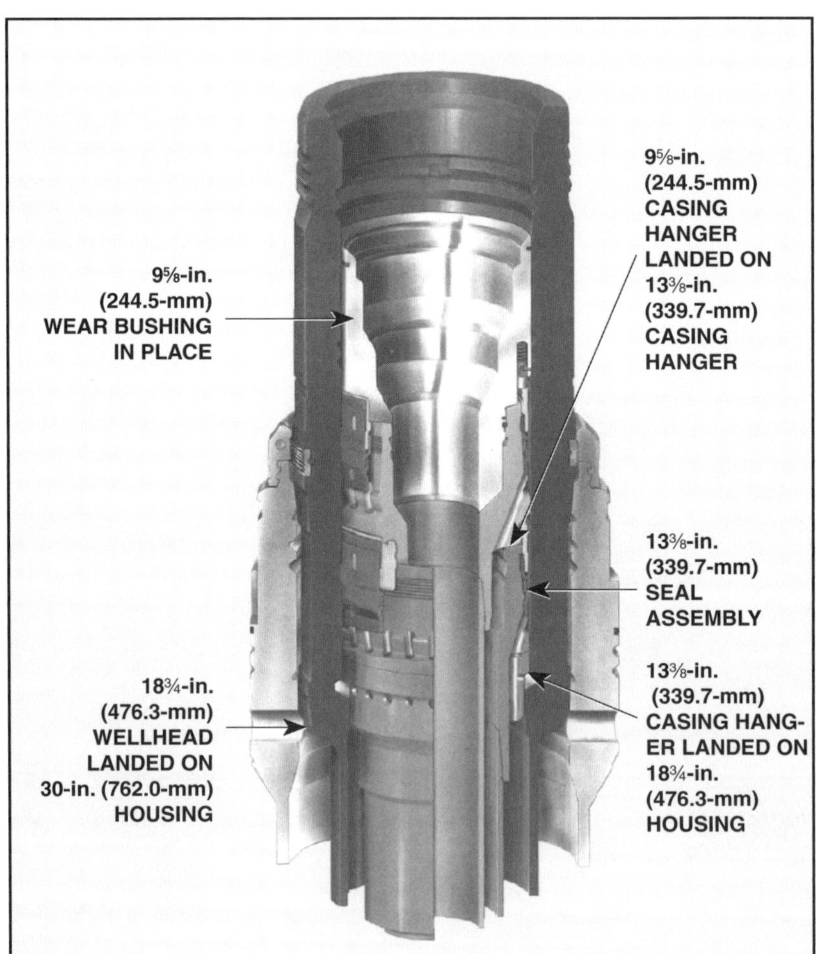

Figure 34. Casing hangers landed on wellheads of various sizes

When well designers select wellheads and casing strings, they have several items to consider. For example, they must—

- ensure that the wellhead selected has the correct pressure rating and ID for the well being drilled.
- ensure that the BOP connector is the correct make and model to mate with the selected wellhead.
- inspect all wellhead sealing areas and ensure that they are in good condition prior to running and cementing in place.
- ensure that the IDs of the wellhead, casing hangers, and wear bushings are correct before the crew runs them. Correct IDs are required so that the correct size of drilling assembly can pass through.
- ensure that the wear bushing is removed prior to running the next casing string.
- ensure that the pressure rating of each casing string is known and documented to avoid overpressuring the casing.

If the operator is to produce the well with a subsea completion, crew members run and land the completion tree on the 18¾-in. (476.3-mm) wellhead after they retrieve the BOP and marine riser.

To summarize—

The subsea wellhead and casing are a system that—

- form the foundation for drilling a successful well.
- link the well to the BOP equipment.
- consists of a structural casing string, which is set and cemented in place to support the BOP assembly at the seabed. This structural casing also supports subsequent casing strings that are run in the well.

The names, hole sizes, and casing sizes for a typical casing program are—

Casing String	Hole Size, in. (mm)	Casing Size, in. (mm)
Structural	36 (914.4)	30 (762.0)
Conductor	26 (660.4)	20 (508.0)
Surface	13⅜ (339.7)	17½ (444.5)
Intermediate	12¼ (311.2)	9⅝ (244.5)
Production	8½ (215.9)	7 (177.8)

The 18¾-in. (476.3-mm) high-pressure wellhead is welded to the last joint of 20-in. (508.0-mm) casing. This wellhead—

- provides the foundation for the BOP stack,
- supports and houses subsequent casing strings, and
- provides a wellbore seal-and-locking arrangement between the surface casing and BOP stack.

The BOP assembly is run on marine riser pipe and landed on the 18¾-in. (476.3-mm) wellhead.

After landing the 13⅜-in. (339.7-mm) casing hanger, subsequent holes are drilled and the casing strings are run and landed in the 18¾-in. (476.3-mm) wellhead. Each casing string has a special hanger, which lands on the previous hanger. For example, the 9⅝-in. (244.5-mm) casing hanger lands in the 13⅜-in. (339.7-mm) hanger and so on.

Subsea BOP Stacks

Subsea BOP stacks for floating drilling rigs are made up of several components (fig. 35). Such stacks are built to enable personnel to control a well under virtually every condition that is likely to be encountered. Crew members run the BOP stack from

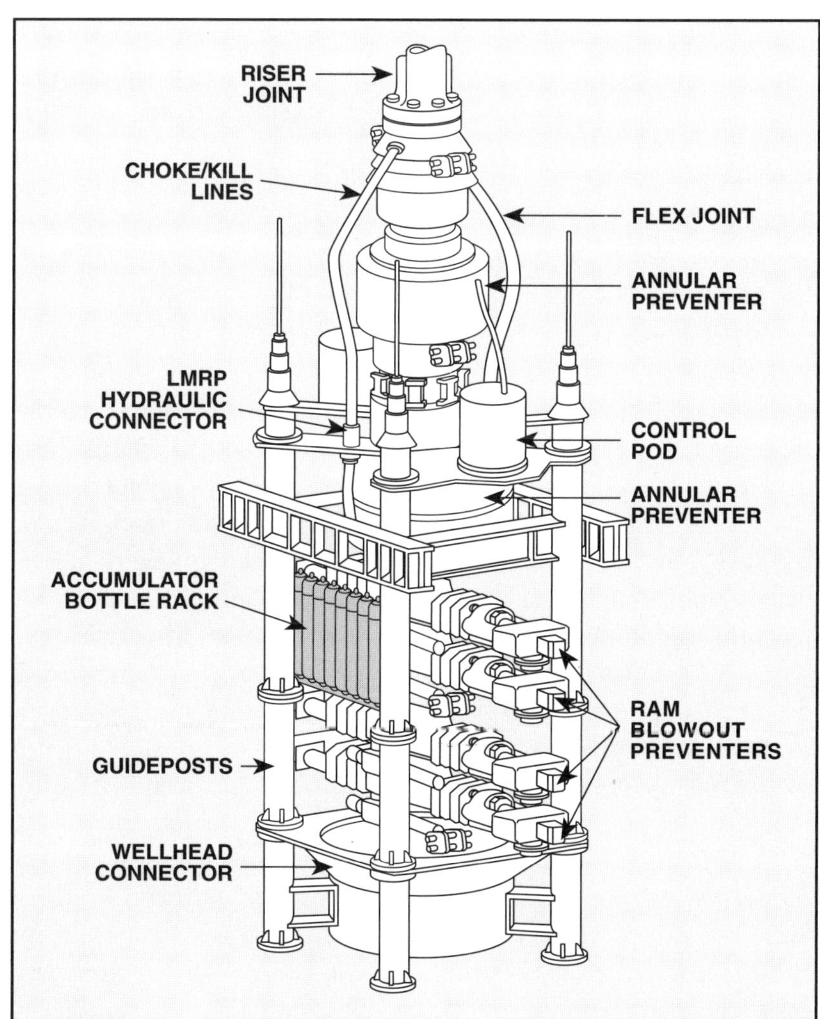

Figure 35. Subsea BOP stack

the surface on the marine drilling riser and attach it to the wellhead on the ocean floor.

Subsea BOP stacks installed at the wellhead for a floating drilling rig serve the same purposes as they do for conventional land drilling: (1) they provide a means of pressure control when a well flow develops; (2) they allow drilling fluid to be circulated and conditioned; (3) they allow the well to be returned to a static condition. Further, the BOP stack must permit personnel to—

- circulate wellbore fluids under pressure for prolonged periods of time.
- hang off the drill pipe in the BOP or at the wellhead.
- seal the well and monitor wellbore pressure.
- shear the drill pipe and disconnect the marine riser from the BOP.
- reconnect to the BOP and monitor wellbore pressure before opening the well.

Subsea BOP stack components are housed within with a steel framework that guides the BOP assembly to the wellhead in conjunction with the rig's guideline system and wellhead guide base. The framework also protects the BOP components from collision damage. Guide bases have standard-size guideposts to accommodate all guideline BOP frames.

The framework is a rigid assembly of pipe and structural steel members with bolted flanges to permit disassembly without cutting. Some parts of the frame's structure may be hinged to permit quick access to BOP-mounted accumulators and other components of the BOP stack. While the main guideposts and structural members are made of relatively heavy steel—they must be very rigid—lighter framing is often used to protect the BOP equipment when the stack is being handled. The guide funnels at the bottom of the framework fit over the guideposts on the wellhead guide base. Guideposts extend upward from the top of the framework to guide the LMRP into place.

On deepwater high-pressure BOP stacks, the framework is designed to prevent high bending forces from being transmitted to the BOP connections.

At water depths of more than 5,000 ft (1,500 m), it is impractical to run guidelines to the seabed. So, guidelineless BOP stack frames are available. They have large-diameter conical funnels, which enable the BOP stack to mate with the wellhead even if some misalignment between the two is present. As the BOP is lowered over the wellhead, the conical funnel automatically aligns the connector to the wellhead. This operation has to be carefully monitored—usually with an ROV camera—to avoid possible damage to the wellhead sealing surface. The same conical funnel arrangement is used between the BOP and LMRP in guidelineless systems.

Manufacturers give BOP stack components pressure ratings of 2K, 3K, 5K, 10K, and 15K, which stands for 2,000, 3,000, 5,000, 10,000, and 15,000 psi (13,790, 20,685, 34,475, 68,950, and 103,425 kPa). Regardless of the BOP stack's pressure rating, once it is installed on the wellhead, the pressure rating of the casing limits the overall system's pressure rating. For testing the BOP on the wellhead, crew members run and land isolation test plugs in the high-pressure wellhead housing. Isolation plugs protect the casing string, which has a lower pressure rating than the BOP stack, from the high pressures required to test the BOPs.

From the bottom up, components in a typical subsea BOP stack consist of—

1. a high-pressure wellhead connector;
2. four ram preventers (in deepwater applications, up to six ram preventers may be used);
3. choke and kill side outlets below each set of ram preventers;
4. high-pressure valves on the choke and kill side outlets;
5. high-pressure choke and kill piping;
6. an annular preventer;
7. a BOP mandrel;
8. BOP-mounted accumulator bottles; and
9. a top receiver plate that houses the choke and kill line connections to the LMRP, the BOP control system receptacles, and LMRP guides.

The LMRP consists of the following components—

1. an LMRP connector,
2. a lower base plate,
3. an annular preventer,
4. LMRP-mounted accumulator bottles,
5. a flex joint,
6. a riser adapter,
7. flexible choke- and kill-line piping,
8. connectors on the choke and kill lines that mate with connections on the BOP, and
9. BOP control pods.

BOP Stacks

A wellhead connector attaches the BOP stack to the wellhead. It is hydraulically operated from the surface and, when actuated, connects the BOP to the wellhead. The connector has the same pressure rating as the ram BOPs, and, when latched, creates a wellbore seal at the wellhead.

Manufacturers supply ram preventers in single units (one ram cavity), double units (two ram cavities). or in triple units (three ram cavities). Crew members install fixed-bore pipe rams, variable-bore pipe rams, or shear rams in the cavities. The ram bodies are connected to each other with flanged, hubbed, or studded connections. Steel gaskets form a metal-to-metal seal between each body. The connections must prevent separation of the bodies that could be caused by wellbore pressure or bending moments exerted through the marine riser and flex joint. Therefore, the connections are *preloaded* as crew members tighten bolts and nuts to connect the components. (Preloaded means that the connections are tightened to final torque under pressure high enough to simulate well pressures.)

The selection of single, double, or triple units is usually based on the handling restrictions on the rig. Generally, multiple units reduce the overall height of the assembly. Both standard and extended double and triple units are available. An extended unit provides more space between the upper ram cavity and the shear ram cavity than a standard unit. Additional space not only allows the drill string to be hung off on a tool joint on the upper rams, but also it provides room for the shear rams to shear the pipe above the tool joint.

The position of the various rams in the stack depends on the operational requirements of a particular well. For example, a well destined for subsea completion may require use of fixed-sized pipe or casing rams or high-capacity shear rams.

Below each ram cavity, side outlets for the choke and kill lines are machined into the ram preventer's body, one on each side of the bore. The side outlets allow the choke and kill lines and their valves to be optimally positioned. Normally, four outlets are connected to the choke and kill system, including two choke outlets and two kill outlets, each of which is positioned below a different set of rams. Should one ram preventer fail while a well is shut in on a kick, another ram preventer can be selected with its corresponding choke or kill line to continue the operation. The positioning of the choke and kill lines also allows the well to be circulated even when sheared pipe is in the hole.

Subsea choke and kill valves are connected to the ram BOP's side outlets and to each other by a flanged, hubbed, or studded connection, with a metal-to-metal seal between the connections.

The connections must prevent separation, which could be caused by wellbore pressure or bending moments exerted through the BOP frame.

Choke and kill lines on a subsea BOP stack are manifolded together on the surface at the choke manifold. Manifolding permits either line to be used as a choke or a kill line. For example, crew members can pump down the drill string and take well returns through either line, or through both lines simultaneously. They can also pump down the choke line and take well returns through the kill line, and vice versa. This arrangement provides additional redundancy to the well control system.

An annular preventer allows the crew to *strip* drill pipe into a pressurized well through the closed preventer. *Stripping* is required if the well kicks while pulling out of the hole. (Stripping means to lower pipe into the hole with a blowout preventer—usually the annular preventer—closed against well pressure.) Stripping allows the crew to get the bit back on bottom to better control the well. One annular BOP is normally positioned in the BOP stack and one in the LMRP. Rig personnel normally use the annular preventer in the LMRP for stripping operations. If this preventer fails, they can bring it to the surface with the LMRP and repair it while leaving the well shut in with the annular preventer in the BOP stack.

For deepwater drilling, manufacturers put side outlets in the annular preventer's body below the packing element. Outlets allow the crew to vent trapped gas below the element. The side outlets are connected to the high-pressure choke or kill lines. These connections allow personnel to vent trapped gas through the lines rather than through the marine riser and diverter system, which may not be able to handle the volumes involved.

Manufacturers also provide side outlets on ram preventers. The high-pressure valves on the ram's side outlets are usually hydraulically-operated gate valves. These valves must have the same pressure rating as the ram preventers. Normally, the ID of the flow ports through the valves is at least 3 in. (76.2 mm). Each side outlet has two valves mounted in series for redundancy. The hydraulic actuators are the fail-safe-to-close type—that is, hydraulic pressure is required to open the valves, but spring force keeps them closed. Thus, the valves automatically close if hydraulic control is lost.

Under certain conditions, however, a fail-safe valve may not close. For example, high-pressure flow through the valve body may prevent the spring from closing the valve when hydraulic power is lost. Consequently, most valves have a backup hydraulic source to ensure full closure of the valves in all situations.

The high-pressure choke and kill lines mounted on the BOP have the same pressure rating as the ram preventers and have a minimum ID of 3 in. (76.2 mm). The piping terminates at the upper BOP receiver plate to either a simple pin arrangement or a special hub. The LMRP has either a corresponding mating box with lip seals that mates with the pin or a hydraulically-operated connector that mates with the special hub. Both these arrangements allow the crew to disconnect the LMRP from the BOP. Either arrangement also seals the choke and kill lines as crew members connect the LMRP to the BOP. In both cases, the LMRP houses the seals because it allows the crew to retrieve and replace the seals should they fail.

The BOP mandrel is connected to the top of the annular preventer in the BOP stack. The mandrel has a mating-and-seal profile to match that of the LMRP connector. The connector normally has the same pressure rating as the LMRP's annular preventer.

The control pod receptacles on the BOP stack's upper receiver plate link the hydraulic control pods on the LMRP to the BOP stack's components. Hydraulic hoses, which are connected to the underside of the receptacles, carry operating fluid to the BOP components. Drilled fluid ports in the receptacles allow operating fluid to flow from the control pod's stab into the hose and to the BOP component. Circular seals, which are mounted in the control pod's stab, seal the fluid ports. If a seal fails, the control pod can be retrieved for seal replacement; or, in the case of deepwater applications, the control pod is retrieved with the LMRP.

A final alignment mechanism on the BOP's upper receiver plate provides accurate alignment of the LMRP to the BOP. Guide funnels, guideposts, and the connector funnel provide initial alignment. Because the LMRP is lowered over the BOP, and at a point about 6 in., or 15 centimetres (cm), before final engagement, a series of tapered pins on the LMRP plate enter receptacles on the BOP plate. The tolerance between the pins and receptacles is sufficient to provide accurate alignment of the choke and kill stabs and the control pod's receptacles.

To provide fast activation of the BOP components, assembly personnel mount accumulator bottles on the BOP stack frame. Accumulator bottles on the stack itself reduce the time the fluid takes to operate the components. (The fluid does not have to travel from bottles on the surface.) As operating the BOPs consumes fluid from the accumulators, the accumulators are recharged through the surface accumulator pumps. In deepwater applications, additional accumulator bottles mounted on the BOP provide fluid for backup operating systems such as acoustic backup, high-pressure shear function, automatic shear, and emergency disconnect systems.

The LMRP (fig. 36) is an interface between the riser system and the BOP stack. It can be disconnected should the need arise.

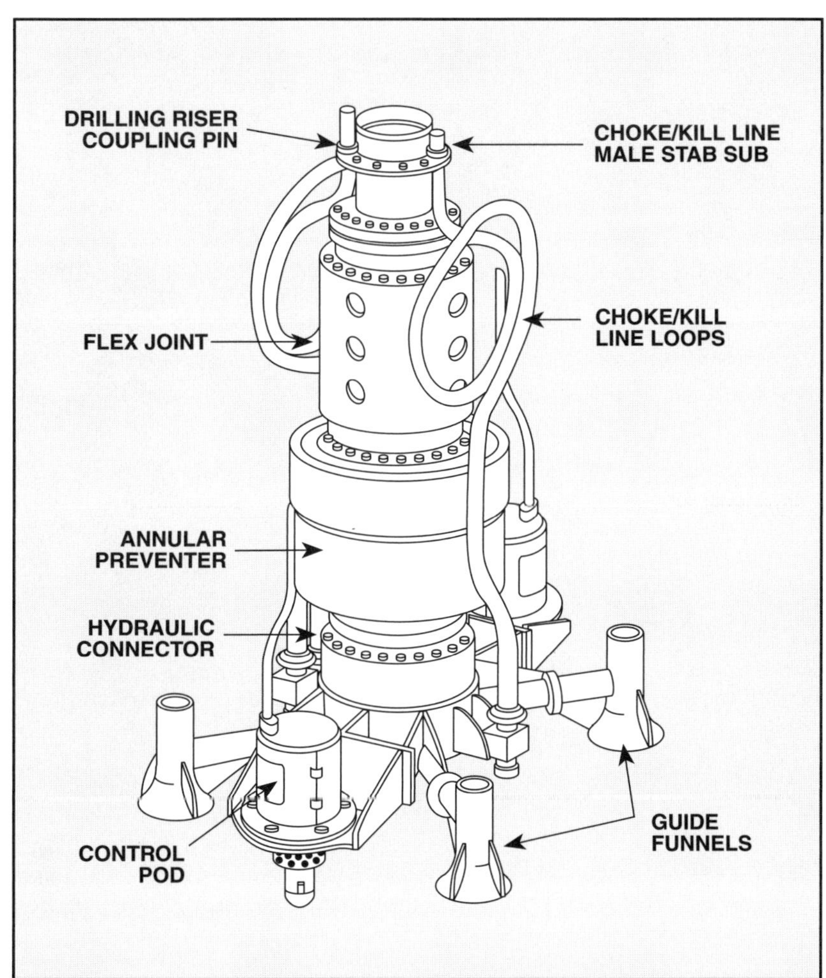

Figure 36. Lower marine riser package (LMRP) (Courtesy of ABB Vetco Gray)

Indeed, the main function of the LMRP is to provide emergency disconnect of the riser from the BOP stack. For example, if the vessel drifts off location, the crew can disconnect the LMRP and riser from the BOP stack on the seafloor (fig. 37).

Figure 37. Vessel being moved off location (A) and LMRP and riser safely disconnected (B)

Hydraulic Connectors

Hydraulic connectors allow the crew to remotely connect the BOP assembly to the wellhead and the LMRP to the BOP. Connector manufacturers include Cooper Cameron, ABB Vetco Gray, and DrilQuip. Connectors are available in several sizes and pressure ranges. Cameron connectors include the Model 70, the HC, and the DWHC. ABB Vetco Gray calls its connector the H4, and DrilQuip terms its connector the DX. All are similar in design and consist of a lower body, an upper body, a cam ring, locking dogs or segments, a seal groove, and a hydraulic system. The ID of the locking dogs or segments has a tapered locking profile that mates with a tapered locking profile on the wellhead's OD. The cam ring is connected to the hydraulic pistons and its ID has a 4-degree taper. When hydraulic locking pressure is applied, the pistons pull the cam ring downward over the locking dogs or segments and the 4-degree taper forces the dogs or segments inward to lock on the wellhead profile.

The combination of locking pressure, the 4-degree taper on the cam ring and locking dogs or segments, and the taper on the wellhead profile pulls the connector's upper hub and the wellhead hub into face-to-face contact. These forces simultaneously energize the stainless steel wellbore gasket. Locking forces also provide the necessary preload to prevent separation at the hubs caused by full wellbore pressure and bending moments.

The hydraulic system is rated to 3,000-psi (21,000-kPa) maximum working pressure and consists of a primary circuit and a secondary circuit. The primary circuit provides primary lock and unlock pressure to operate the connector. The secondary circuit, which is separate from the primary circuit, provides additional unlocking force that may be required to overcome high-friction forces generated between the tapers on the cam ring and the locking dogs or segments. Combined primary and secondary unlock pressure is about 1.8 times higher than primary locking force alone.

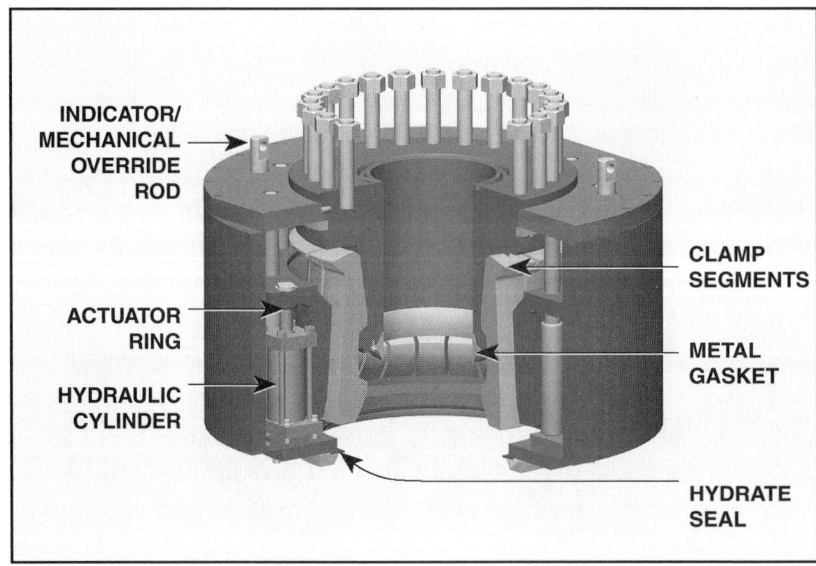

Figure 38. Cameron Model 70 connector (Courtesy of Cameron)

The Cameron Model 70 connector (fig. 38) uses hydraulic cylinders in the connector's lower body to operate an actuator ring. The connector has nine cylinders, six of which are primary and three of which are secondary. The six primary pistons are connected to the actuator ring through connecting rods. The pistons can pull the ring downward to lock, or push it upward to unlock. The secondary pistons, however, can only push the ring upward to unlock. Model 70 connectors are available with pressure ratings as high as 10,000 psi (70,000 kPa).

The Cameron HC connector (fig. 39) is similar to the Model 70, but, instead of individual pistons, the HC has a large annular piston that completely surrounds the locking segments. The additional locking force available through the annular piston increases the pressure rating to 15,000 psi (105,000 kPa). The HC connector is also more resistant to bending moments than the Model 70. The HC's larger unlocking area—905 in^2 (5,837 cm^2) versus 732 in^2 (4,721 cm^2) in the Model 70—provides extra unlocking force to overcome friction forces. Cameron also provides a secondary unlock piston on the HC, but it does not provide additional unlocking force. Instead, it serves as a backup if the seals on the annular piston fail.

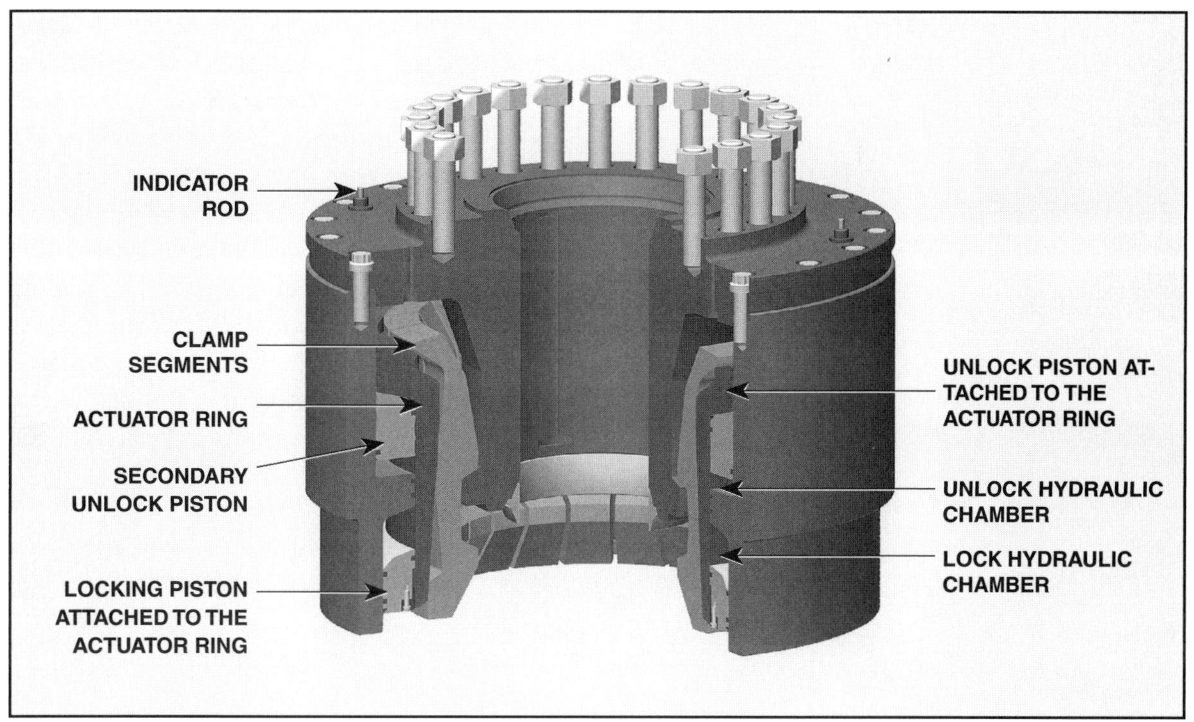

Figure 39. Cameron HC connector (Courtesy of Cameron)

In the unlocked position, the Cameron Model 70 and HC connector locking segments provide a 30-degree lead-in taper for aligning the connector to the wellhead. This lead-in taper, combined with a relatively shallow insert depth of 12½ in. (318 mm) over the wellhead, makes these connectors ideal for use on the LMRP, where the crew may have to release it from a high angle.

Cameron also manufactures a miniature connector (miniconnector) that connects the choke and kill lines at the BOP-to-LMRP connection. Miniconnectors operate in the same way as HC connectors. The miniconnector has a pressure rating of 15,000 psi (105,000 kPa) and a 3-in. (76.2-mm) ID.

The ABB Vetco Gray H4 connector was originally built much like the Cameron Model 70 connector. Only half of the pistons—the primary pistons—were connected to the cam ring to pull downward and provide locking force. The other half of the pistons—the secondary pistons—could apply upward unlocking force only. It was found that the preload available with this set up was insufficient. Today, both the primary and secondary pistons in H4 connectors are connected to the cam ring to provide locking and unlocking force (fig. 40). The additional unlocking force is provided by the larger piston area on the unlock side compared to the lock side, which consists of the piston area minus the rod area. The number of pistons in the H4 connector is related to the size and pressure rating of the connector.

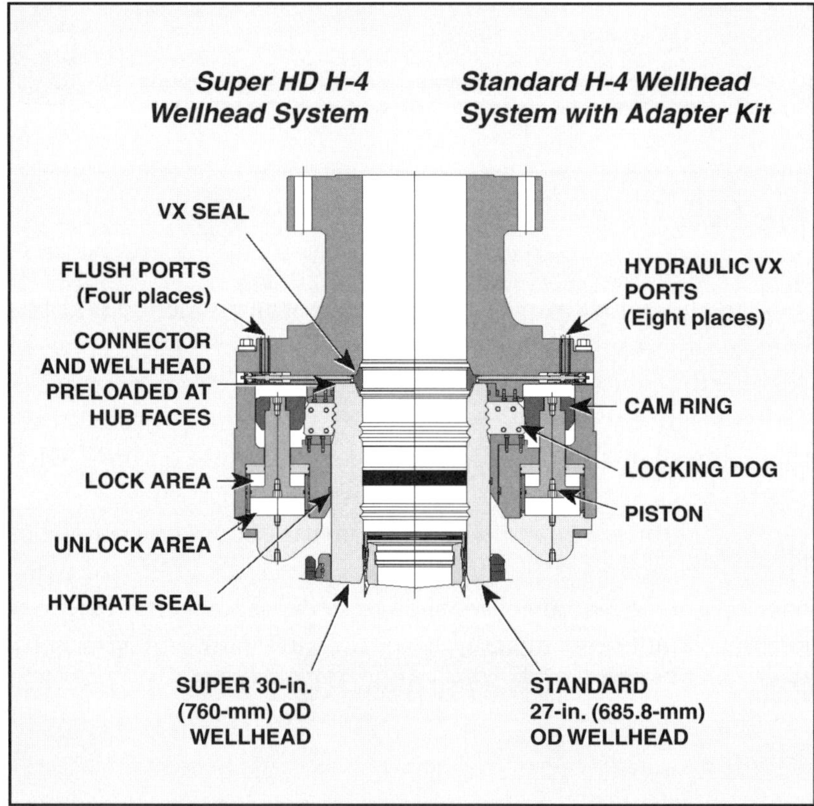

Figure 40. ABB Vetco H4 connector (Courtesy of ABB Vetco Gray)

Tapered cylinder heads mounted on the bottom of the H4's lower body provide connector lead in and final alignment. H4 connectors with ratings of 10,000 and 15,000 psi (70,000 and 105,000 kPa) generally fit 27½-in. (698.5-mm) wellheads. However, some are sized to fit 32-in. (812.8-mm) wellheads. The ability to fit such large wellheads makes it difficult to release the connector if it is misaligned and creates an angle.

Hydraulically operated, spring-loaded pins hold the stainless steel connector gasket that seals the wellbore in place. The pins protrude from the connector body and fit under a support shoulder on the gasket. Hydraulic force releases the pins and spring force retains the gasket. Take care to ensure that the pins and springs are in good condition; otherwise, the gasket could release from the connector as the BOP is being run.

Hydrates are hydrocarbon and water compounds that resemble snow or ice. They can form at the very low temperatures found at extreme water depths. So, in deepwater operations, hydrates can occur in connectors and affect their operation. For example, hydrates can prevent the connector from unlocking if they form around the actuator ring and locking dogs. Therefore, all deepwater connectors should have hydrate seals and be fitted with glycol injection ports. The hydrate seal is placed in a groove in the lower body and forms a seal with the outside of the wellhead. The injection ports are located in the upper body and allow glycol to be injected for hydrate prevention.

Important points to note about hydraulic connectors include—

- replace the wellbore stainless steel gasket each time the BOP is run to the wellhead.
- ensure that the new gasket is securely retained in the connector prior to running the BOP.
- ensure that the manufacturer's recommended lubricant is used on the cam ring and the locking dogs or segments to avoid excessive friction build up.

Ram Preventers

Cooper Cameron, Varco Shaffer, and Hydril are the main manufacturers of ram preventers. Regardless of the manufacturer, ram preventers operate in the same way and serve the same purpose. That is, they close around drill pipe or on open hole to seal the hole. Pipe ram preventers seal the annulus between the drill pipe and the wellbore below. Blind ram preventers close in the top of the open hole. However, blind ram preventers are not normally installed in a subsea stack. Instead, blind-shear rams are used. Blind-shear rams not only seal on open hole, but also can cut drill pipe and form a seal to shut in the well.

Ram preventers are available in pressure ratings of 2,000, 3,000, 5,000, 10,000 and 15,000 psi (14,000, 21,000, 35,000, 70,000, and 105,000 kPa). (At least one manufacturer makes a 20,000-psi or 140,000-kPa ram preventer. However, availability is limited.) Ram preventer sizes, which are based on the diameter of the bore through the middle of the preventer, range from 7 to 21¼ in. (177.8 to 539.8 mm). Most offshore drilling rigs use 18¾ in. (476.3 mm) ram preventers with pressure ratings of either 10,000 or 15,000 psi (70,000 or 105,000 kPa).

Pipe Ram Preventers

Two cavities on opposite sides of the preventer's body house the steel ram blocks. In pipe ram preventers, the ram blocks contain the front packer and top seal that create a seal between the drill pipe, ram block, and cavity. In the open position, the ram blocks are clear of the wellbore and allow passage of full-bore tools. The ram blocks are connected to hydraulic pistons that push the two ram blocks into the wellbore simultaneously to form a seal against the drill pipe. As the ram blocks come together around the drill pipe, the front packers make contact with the pipe and with each other (fig. 41). The force exerted by the hydraulic pressure creates compression in the packers that is sufficient to overcome wellbore pressure, which would tend to separate them and break the seal. So, the amount of wellbore pressure that can be sealed against is directly related to the amount of closing pressure applied.

The heavy steel ram blocks contact the bottom of the ram cavity. Consequently, as the rams are closed and opened, wear occurs on the two steel surfaces. Eventually the wear becomes great enough to impair the rams' sealing capability. To combat wear, manufacturers install replaceable wear plates in the cavities or provide replaceable wear pads for the ram blocks. Crew members must regularly measure for the amount of wear to prevent failure of the BOP.

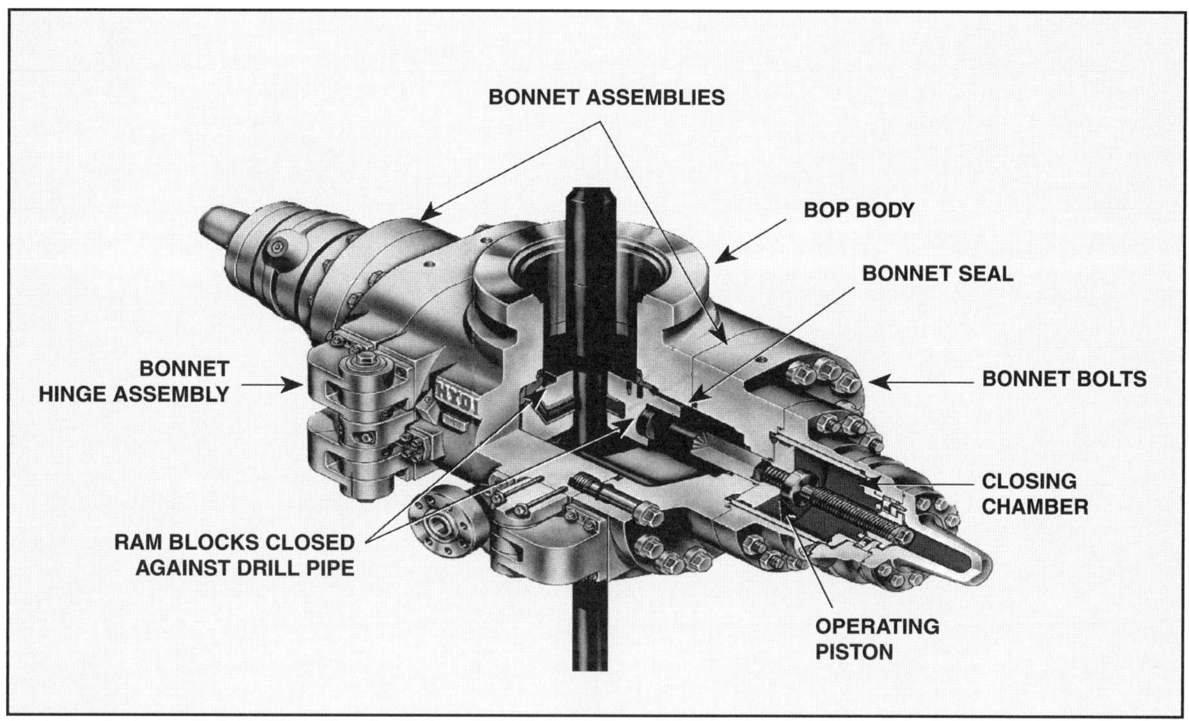

Figure 41. Pipe ram preventer closed on drill pipe (Courtesy of Hydril)

Ram packers and top seals are manufactured from a synthetic rubber compound that is molded to upper and lower steel antiextrusion plates. The steel antiextrusion plates prevent the rubber from being forced out of the small area between the drill pipe and the ram block when wellbore pressure is applied. Further, they feed reserve rubber into sealing contact with the drill pipe as rubber is lost because of wear. The reserve of rubber sits behind the support plates in the ram block. As the rubber wears, the ram blocks move closer to the drill pipe; eventually, the support plates in the opposing rams make contact. Hydraulic closing force then pushes the support plates back into the ram block. In turn, the support plates force the reserve rubber out of the ram block and into contact with the pipe. The top seal completes the sealing arrangement between the ram block and the cavity.

Ram blocks retain pressure from below only; therefore, pressure should never be applied from above. In ram preventers, wellbore pressure helps keep the rams closed. Thus, the higher the wellbore pressure, the higher is the closing force on the rams. When the rams are closed, wellbore pressure enters the cavity behind the rams where it provides an additional closing force. Consequently, rams should never be opened when wellbore pressure is applied to the preventer. Opening a closed ram preventer can damage the ram rubbers, ram blocks, cavities, and the hydraulic pistons.

Manufacturers provide ram blocks in all sizes of drill pipe and casing normally run through the BOP (fig. 42). Fixed-size ram blocks can close and seal only on the size of pipe for which they are designed. For example, fixed 5½-in. (139.7-mm) rams can only close and seal on 5½-in. (139.7-mm) drill pipe, and fixed 7-in. (177.8-mm) rams can only close and seal on 7-in. (177.8-mm) pipe. Closing fixed rams on pipe of a size different from the size of the ram blocks can seriously damage the ram packers, ram blocks, or both.

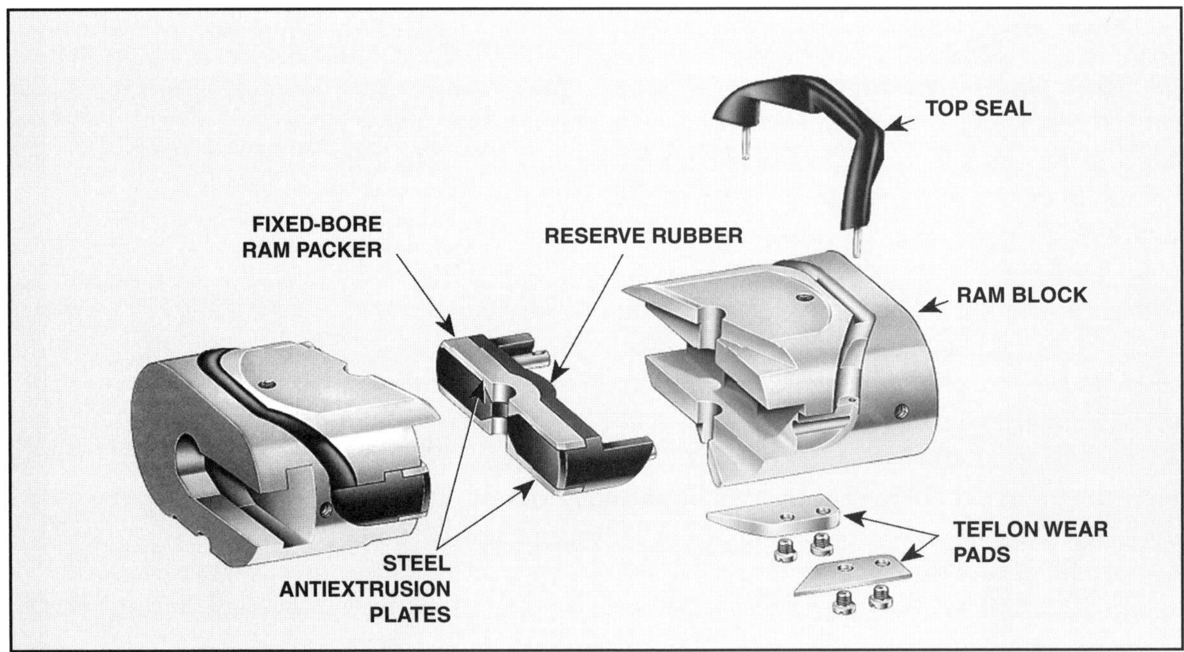

Figure 42. Ram blocks and fixed-bore packers in a pipe ram preventer (Courtesy of Cameron)

Fixed-bore ram blocks are available that can support the load of the drill string when it is necessary to *hang off* the drill string. (Hanging off the drill string means to close the pipe rams just below a tool joint. When weight on the drill string is slacked off, the closed ram blocks support the drill string. Hanging off may be required when it is necessary to disconnect from the wellhead and move the rig—as may be the case when the rig is threatened by a hurricane.) Ram blocks designed for hang off can support up to 600,000 lb or 272,161 kilograms (kg) of drill string load through the string's tool joint. A hardened center area in hang-off ram blocks cuts into the tool joint's taper when the blocks are closed. The hardened center area creates a circular shoulder in the tool joint to help support the weight and to prevent the taper from forcing the rams open.

Variable bore rams can close and seal on a range of pipe sizes—for example, from 3½- to 7-in. (88.9- to 24.5-mm) pipe (fig. 43). Specially designed antiextrusion plates force a large reserve of rubber into contact with the smaller sizes of pipe when wellbore pressure is applied. The antiextrusion plates also support the excess rubber. Variable bore rams can hang off the drill string, but they cannot support loads as heavy as fixed-bore rams.

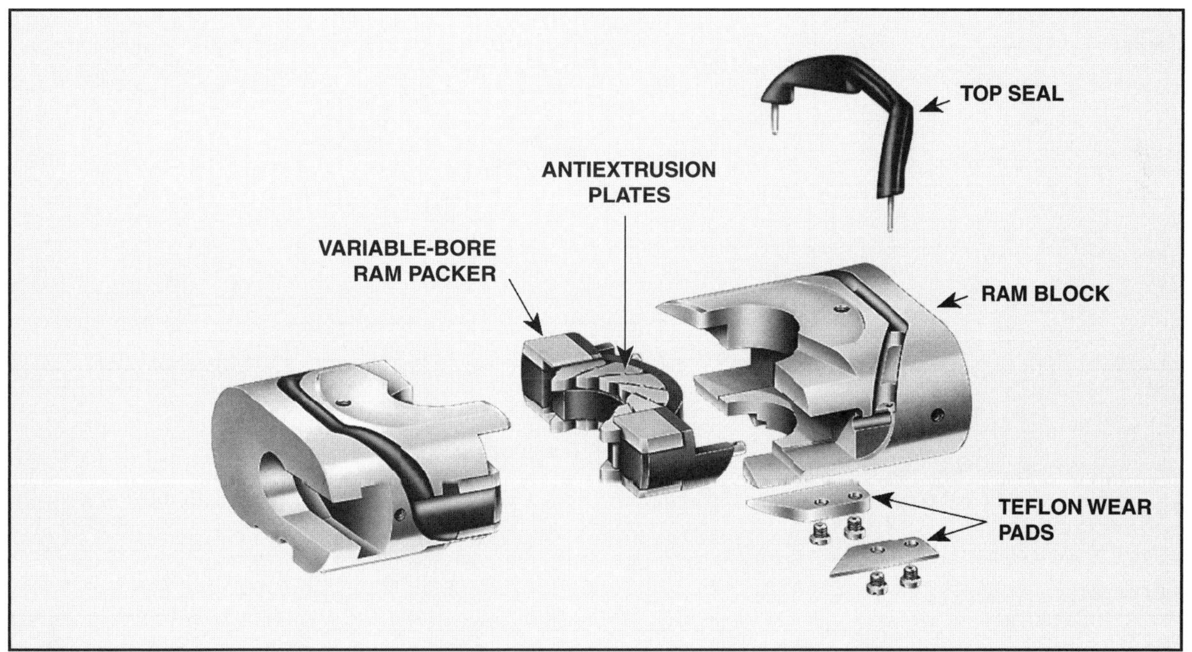

Figure 43. Variable bore packers for a pipe ram preventer (Courtesy of Cameron)

Blind-Shear Ram Preventers

Blind-shear ram preventers can be used as a blind ram preventer, which can seal (close in the well) when no pipe is in the BOP; and as a shear ram preventer, which can cut, or shear, the pipe in the BOP. The drill pipe may need to be sheared to disconnect and move the rig in an emergency. Shearing may also be necessary when a disconnect has accidentally happened. And, some well-control situations require that the drill pipe be sheared.

Shear rams normally require higher closing pressures to shear the pipe than the pipe rams require to create a seal on the drill pipe. Indeed, some rigs have two BOP control hookups: one is a standard hookup for closing the shear rams to seal the well with no pipe in the BOP; the other is a high-pressure control hookup for shearing pipe. All manufacturers supply various grades of blind-shear ram that have different shearing capabilities (fig. 44). Rig supervisors must be aware of the capabilities of the installed blind-shear rams and they must ensure that sufficient hydraulic pressure is available to carry out the shear operation.

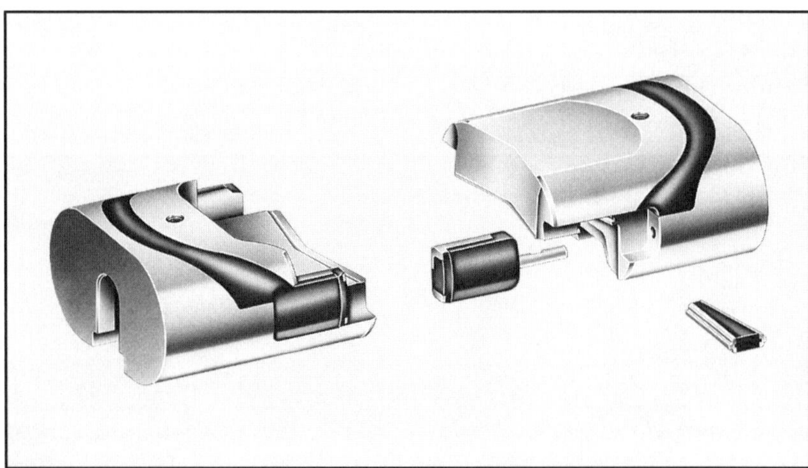

Figure 44. Blind-shear ram blocks (Courtesy of Cameron)

Casing Shear Ram Preventers

Nonsealing casing shear rams can be used in a BOP stack in addition to the blind-shear rams. Casing shear rams are typically high-capacity shear rams capable of shearing drill collars and casing strings. Usually, they are placed below the blind-shear rams in the BOP stack. When casing shear rams are installed and it is necessary to shear pipe, the casing shear rams shear the pipe and the blind-shear rams above the casing rams are closed to seal the well.

Operating and Maintenance Considerations

Important points to keep in mind about ram preventers include—

- Install the correct size rams for the drill pipe in use.
- Record the type and size of the rams in each BOP cavity.
- Accurately record the distance to each ram cavity from the rotary table.
- Do not apply pressure above the ram blocks.
- Do not apply opening pressure to rams when wellbore pressure is applied.
- Install the rams for drill pipe hang off in the appropriate cavities.
- Know the capabilities of the shear rams installed.
- Regularly measure the wear between the ram block and the cavity.
- Regularly inspect the ram packers and top seals.
- Because ram packers and top seals have various temperature ratings, use the correct ram packers and top seals for well conditions.
- Do not close pipe rams or variable bore rams without pipe in the BOP; otherwise, the ram packers will be damaged.

Ram Preventer Operating System

The operating system of a ram preventer consists of a hydraulic cylinder and piston, which are enclosed in a housing called a bonnet. Each ram cavity has one cylinder assembly, or bonnet. The piston is connected to the ram block (see fig. 41). The bonnets are held to the body by several bolts, which are removed to gain access to the ram cavities and ram blocks for inspection and maintenance. After the bolts are removed, the bonnets are opened either by rotating them around a hinge pin, or by hydraulic pressure. Once inspection and maintenance are complete, the bonnet-to-body wellbore seals are replaced, the bonnets are closed against the body, and the bolts torqued to the correct value. The torque applied to these bolts is important to the wellbore sealing capability of the BOP. As wellbore pressure is applied, it contacts the face of the bonnet and acts to open it. If the bolts are not correctly torqued, the pressure can push the bonnet away enough to create a leak at the bonnet-to-body seal.

The maximum hydraulic operating pressure of ram preventers is 3,000 psi (20,500 kPa), and the normal operating pressure is around 1,500 psi (10,500 kPa). To close a set of rams, hydraulic pressure is supplied to a port on the BOP body where it enters internally drilled fluid passages. The passages simultaneously direct the fluid to the close side of the pistons in both bonnets. The pistons are pushed towards the center of the preventer by the fluid pressure, which, in turn, pushes the ram blocks towards the center. As the fluid flows into the close side of the cylinder, the fluid in the open side is pushed out and is allowed to vent at the control pod. The open sequence is the reverse of the close sequence.

The operating hydraulic piston encounters wellbore pressure as it pushes the ram block. Thus, wellbore pressure must be overcome. For this reason, a closing ratio is built into the design of the hydraulic piston. The closing ratio represents the difference in area between the part of the piston that is affected by wellbore pressure, and the part of the piston that closing pressure acts on. The closing ratio is normally around 7 to 1 on a 10,000-psi (70,000-kPa) BOP, but can be higher on a 15,000-psi (105,000-kPa) BOP. To calculate the required closing pressure for a given wellbore pressure, divide the wellbore pressure by the closing ratio.

For example, to determine the closing pressure required where wellbore pressure is 10,000 psi (70,000 kPa) and the closing ratio is 7 to 1, simply divide 10,000 psi (70,000 kPa) by 7. The answer is 1,429 psi (10,000 kPa). Being aware of the closing ratio is important to the safe operation of the BOP.

Ram Locking Systems

All ram preventers have a means of mechanically locking the rams in the closed position to ensure that the rams remain closed with the packers fully energized even if closing pressure is lost. Suppose, for example, that the LMRP must be disconnected from the BOP stack and the rams must remain closed. Once the LMRP is detached from the stack, no hydraulic pressure is supplied to the BOP components. However, if the rams are mechanically locked, they remain closed without hydraulic operating pressure.

BOP manufacturers provide a means of mechanical ram locking that is either actuated through the control system after the rams have been closed, or is an integral part of the hydraulic bonnets and actuates automatically when the rams are closed.

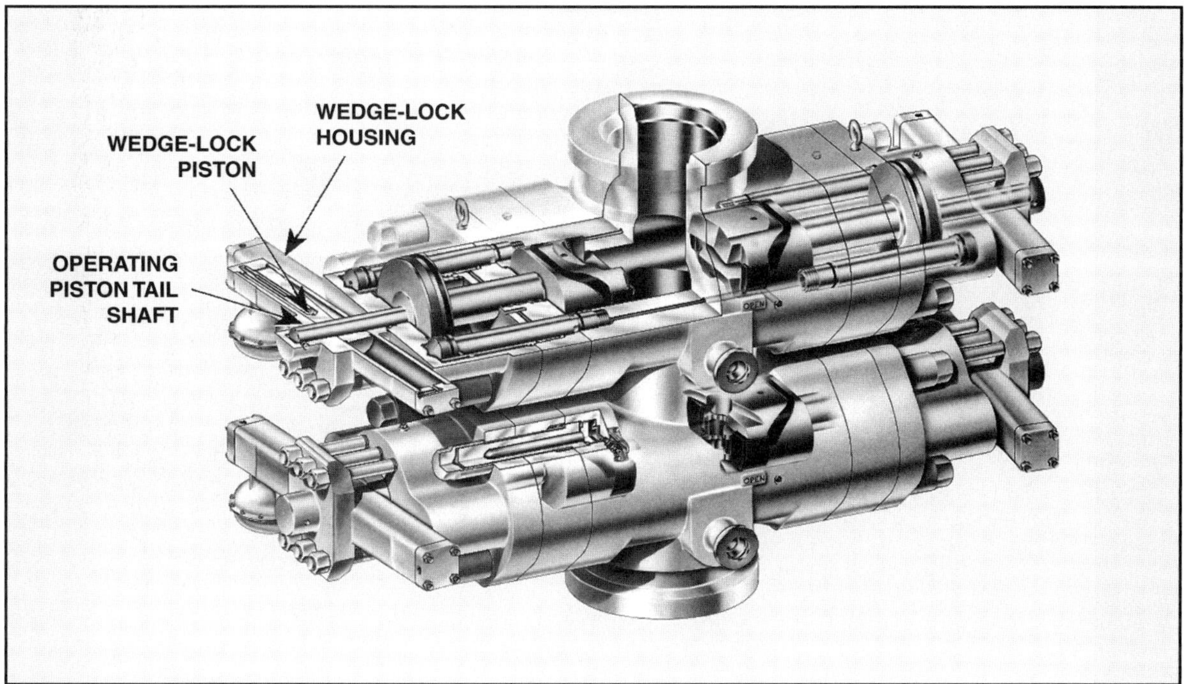

Figure 45. Cameron Wedge-Lock system (Courtesy of Cameron)

One example is the Cameron Wedge-Lock system (fig. 45). This locking system features a wedge-shaped piston that is mounted in its own hydraulic housing on the back of each bonnet at a 90-degree angle to the operating piston's tail shaft. Once the rams are closed, the Wedge-Lock pistons are hydraulically operated to move behind the operating piston's tail shaft and wedge it in place. Until operating pressure is applied to unlock the Wedge-Lock piston, the rams are mechanically wedged closed. An additional hydraulic circuit is required to operate the Wedge-Lock system.

When the rams are in the open position, the operating piston's tail shaft passes through an orifice in the Wedge-Lock's piston. When the rams are closed, the operating piston's tail shaft clears the Wedge-Lock's piston, which can then be hydraulically activated to slide behind the operating piston's tail shaft, thereby wedging it closed.

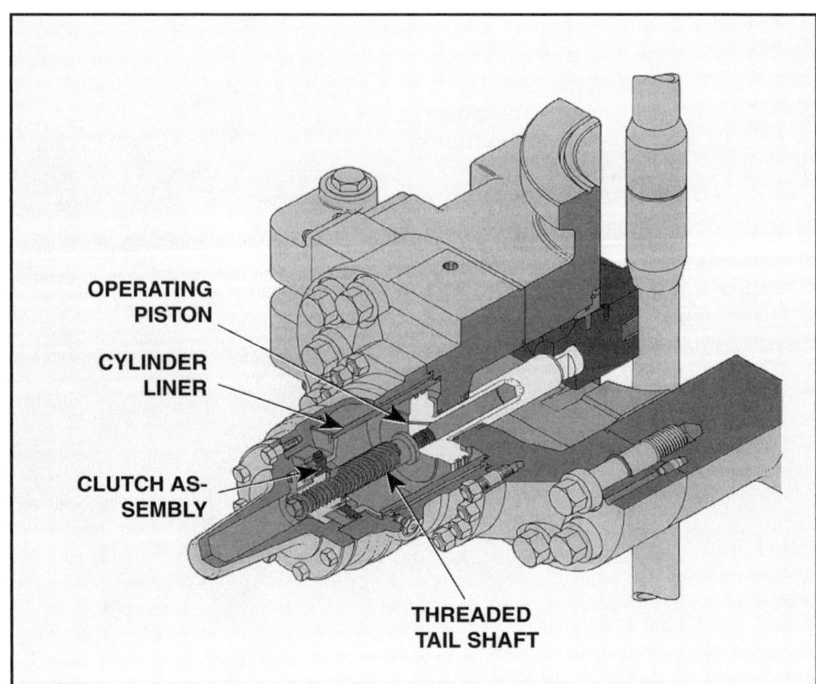

Figure 46. Hydril MPL (multiposition ram lock) (Courtesy of Hydril)

The Hydril MPL (multiposition lock) is an integral part of the hydraulic bonnet operating system, and actuates each time the rams are closed (fig. 46). This system consists of a threaded tail shaft on the operating piston, and two circular clutch plates (fig. 47). One of the clutch plates is connected to the tail shaft through a threaded nut that rotates freely around the thread as the piston is moved toward the closed position. Figure 48 shows the fixed clutch plate. The entire clutch assembly is held in the bonnet housing, and cannot move laterally in relation to the bonnet. The clutch plate surfaces are forced together by spring force and have serrated teeth that allow the rotating clutch plate to rotate in only one direction as the piston moves to the closed position. The teeth engagement prevents rotation if the piston attempts to move toward the open position. Thus, the piston cannot move because the clutch connected to the thread cannot rotate. As the rotating clutch plate rotates freely when the operating piston is moving towards the closed position, the serrated teeth on the two clutch plates jump over each other against the spring force that is keeping them together. When the rams are fully closed, they are locked instantaneously as the spring force engages the teeth on the clutch plate.

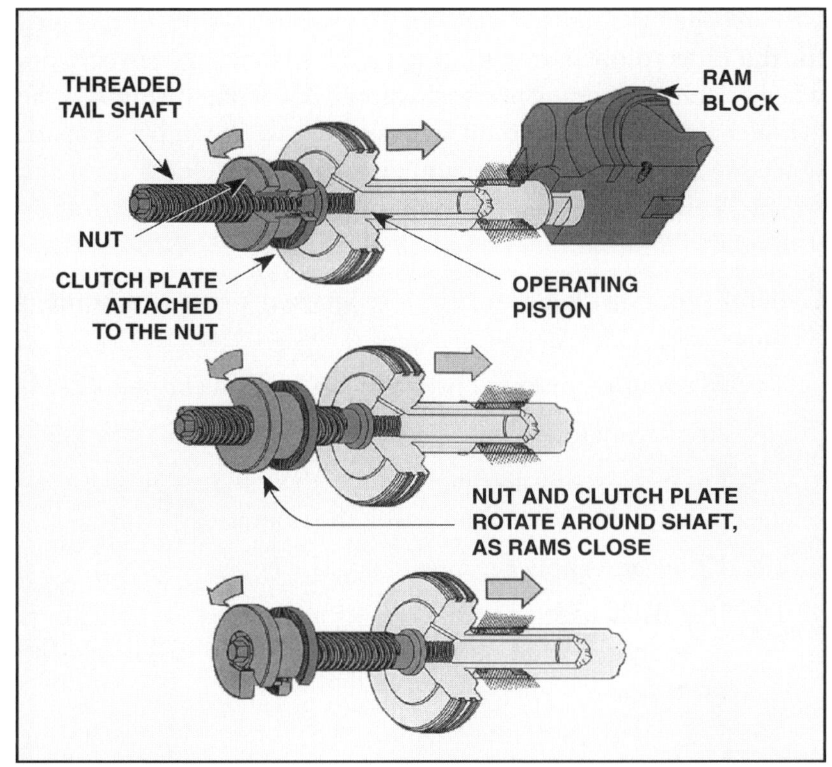

Figure 47. Tail shaft and clutch plates in a Hydril ram preventer (Courtesy of Hydril)

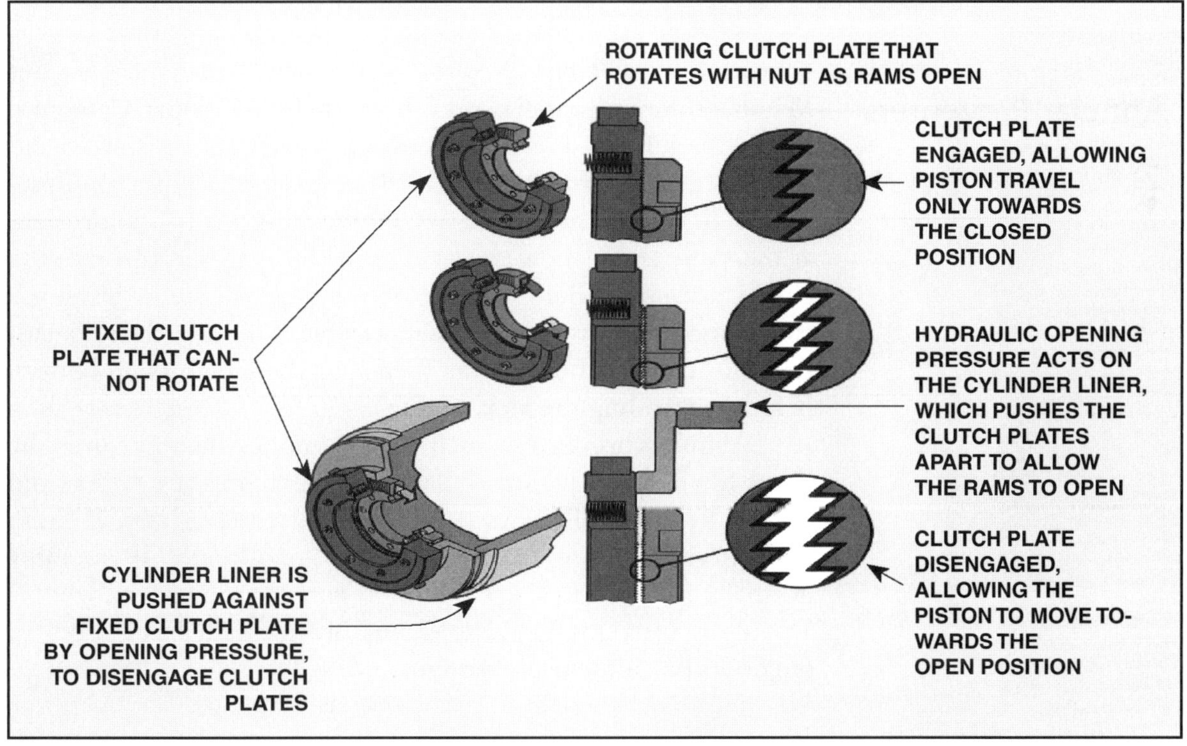

Figure 48. Clutch plate operation in Hydril MPL locking system (Courtesy of Hydril)

To open the rams, the clutch plates must first be disengaged for the rams to open, because the clutch plate cannot rotate. So, when hydraulic opening pressure is applied, the pressure moves the cylinder liner a small amount to push the two clutch plates apart. Once the clutch plates are separated, the rotating clutch can again rotate freely, and the piston moves to the open position.

Ram Bonnet Considerations

General points to keep in mind about ram preventer bonnets include—

- Maximum operating pressure is 3,000 psi (21,000 kPa).
- Approximate working pressure is 1,500 psi (10,500 kPa).
- Do not attempt to open rams with wellbore pressure applied; doing so could damage the operating piston.
- Know and apply the closing ratio.
- Regularly maintain the bonnets to ensure that no leakage of hydraulic fluid occurs, which results in loss of closing pressure.
- Use clean BOP fluid to avoid damaging the operating cylinders and pistons.

Annular Preventers

Manufacturers of annular preventers include Cooper Cameron, Varco Shaffer, and Hydril. The three preventers are similar and perform the same function in the BOP assembly. The main function of an annular preventer is to close and seal the wellbore and, at the same time, allow the drill stem to be moved through the closed preventer. Being able to move the drill stem through a closed annular preventer makes it possible to close in the well with the drill string off bottom and then strip the pipe back to bottom before controlling the kick.

Annular preventers such as the model manufactured by Varco Shaffer (fig. 49) are available in several pressure ratings and sizes. They normally have a lower pressure rating than the ram preventers. For example, a 10,000-psi (70,000-kPa) BOP assembly normally has 5,000-psi (35,000-kPa) annular preventers, while a 15,000-psi (105,000-kPa) BOP assembly normally has 10,000-psi (70,000-kPa) annular preventers.

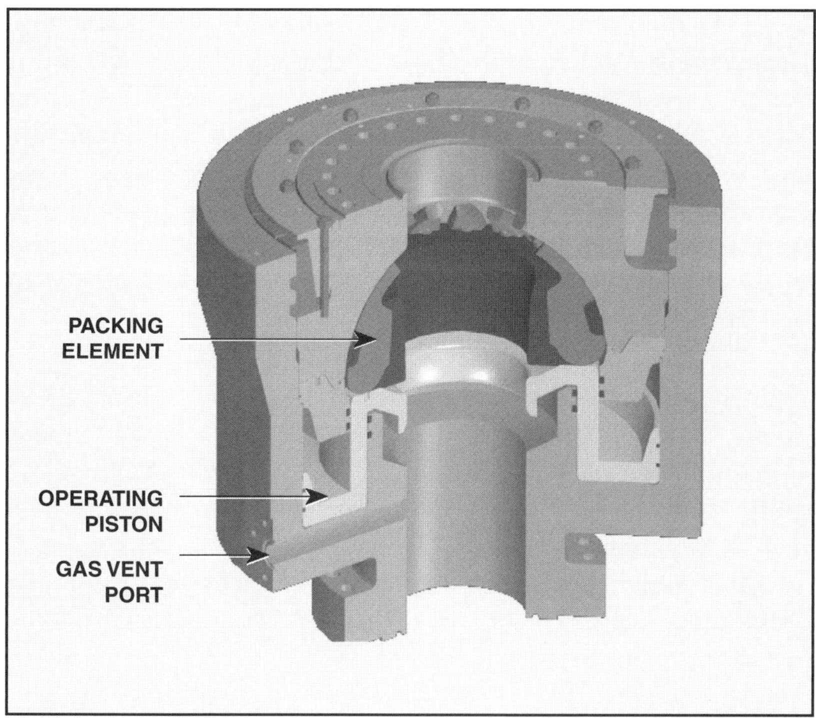

Figure 49. Varco Shaffer annular BOP (Courtesy of Varco Shaffer)

Unlike ram preventers, annular preventers can seal around most objects in the wellbore, such as drill collars, casing, and drill pipe. They are also capable of sealing an open wellbore. However, closing on open hole significantly shortens the packing element's life, so this operation is not recommended unless absolutely necessary.

The main feature of the annular preventer is the packing element, which is a rubber or an elastomer ring molded around several metal support fingers. The element's ID is slightly larger than the BOP's ID to allow full-bore tools to pass through. Three types of material are used in the manufacture of annular packing elements: (1) natural rubber, (2) nitrile synthetic rubber, and (3) neoprene synthetic rubber.

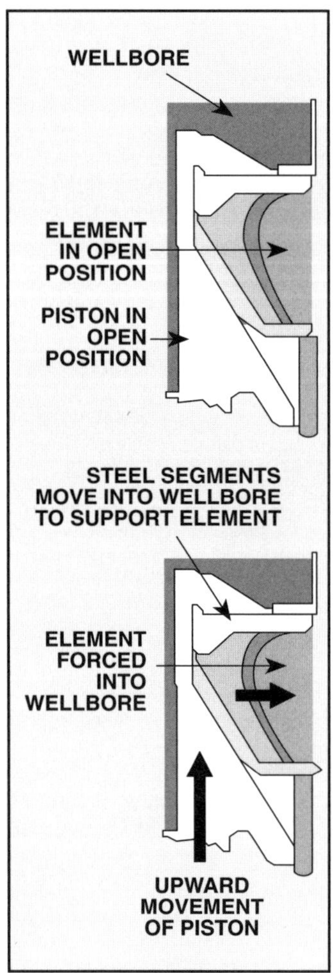

Figure 50. In an annular preventer, upward movement of the piston forces the packing element inward to seal on pipe (Courtesy of Hydril)

Natural rubber elements have superior wear resistance. However, they should be used only with water-based muds and at temperatures ranging from –30 to 225°F (–34 to 107°C). Contact with oil-based mud greatly reduces the life of natural rubber elements. Packing elements manufactured from nitrile synthetic rubber are recommended for use in oil-based mud and at temperatures ranging from 30 to 180°F (–1 to 82°C). Packing elements manufactured from neoprene are recommended for use in oil-based mud at temperatures ranging from –30 to 170°F (–34 to 77°C).

Selection of the correct material for the well conditions is vitally important to the performance of the packing element. The element can fail prematurely if the material from which it is made does not match the conditions in the well. Selecting the material from which the packing element is made should be made on a well-by-well basis.

In annular preventers, the packing element sits on top of a piston that is housed within the preventer's lower body. Hydraulic pressure applied to the closing chamber below the piston forces the piston upward and, as the piston rises, the packing element is forced against the upper body. The packing element, upper body, and piston are designed so that the upward force exerted by the piston squeezes the ID of the packing element into the wellbore to create a seal around the pipe in the bore (fig. 50).

The metal fingers molded to the packing element support the element as it closes, and prevent the packing element from extruding when wellbore pressure is applied from below. Application of opening pressure to the hydraulic chamber above the piston forces the piston downward and allows the elasticity of the packing element to force itself open. Only the elastic properties of the rubber in the element pull the element fully open.

Because the collapse resistance of most casing strings is relatively low, crew members must use a relatively low pressure to close an annular preventer on casing. Normal practice is to close the preventer with minimum closing pressure, and then increase the closing pressure sufficiently to maintain a seal as wellbore pressure increases. The manufacturer provides a closing pressure chart for each casing size.

During drill string stripping operations, the closed packing element must be protected from damage caused by closing pressure surges. Tool joints passing through the element causes pressure surges. Surges lead to rapid wear on the ID of the element. To absorb surges, a 10-gal or 38-litre (L) accumulator bottle, precharged with nitrogen to approximately 500 psi plus 0.450 psi/ft (3,500 kPa plus 10.2 kPa/m) of water depth is mounted in the closing circuit as close to the closing port as possible. When a tool joint passes through the element, closing fluid flows to the accumulator. This action stops surges and prevents rapid wear of the element.

Several lip seals prevent wellbore fluids from entering the hydraulic system. Seal failure can lead to severe damage of the preventer. High pressure—up to 10,000 psi (70,000 kPa) for example—could enter a hydraulic chamber rated to only 3,000 psi (21,000 kPa). To prevent such an occurrence, Cameron Type D (fig. 51) and DL preventers have one lip seal that faces the wellbore, and another lip seal that faces the hydraulic chamber. These two seals create a hydraulic seal. Between the two seals, a passage is drilled through the body, which vents any wellbore fluid leaks to the atmosphere, thereby protecting the hydraulic chamber.

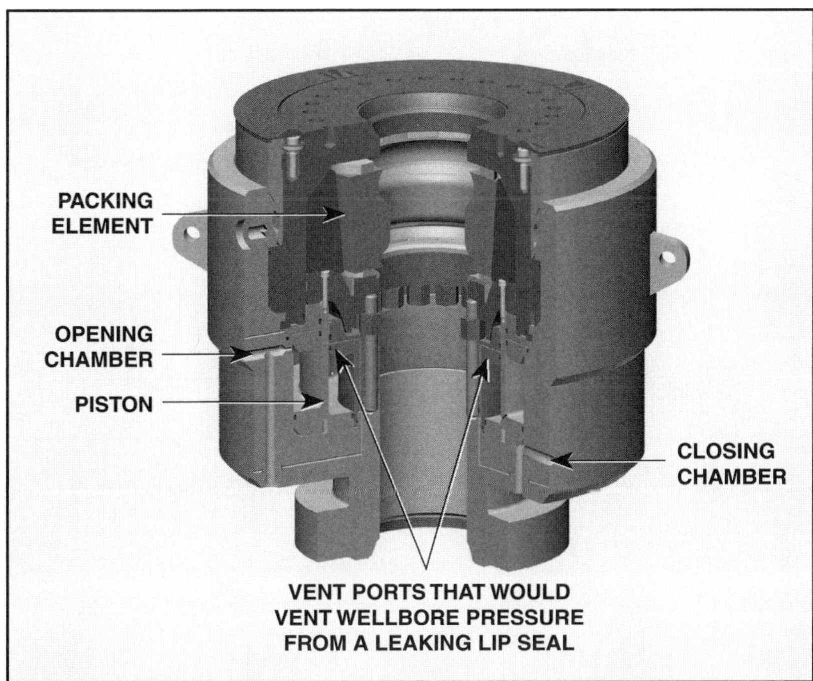

Figure 51. Cameron type D annular preventer (Courtesy of Cameron)

Except for the Hydril GL, annular preventers are sensitive to the water depth in which they are employed. That is, the mud column's hydrostatic pressure acts on the piston to create an opening force against it. When closing a subsea annular, the mud column hydrostatic pressure has to be overcome by the closing force exerted by the closing pressure. Normally, the calculated mud column's hydrostatic pressure is added to the closing pressure required on surface.

The Hydril GL preventer (fig. 52) has an additional hydraulic chamber called the balance, or secondary, chamber. The area of the balance chamber equals the area of the piston exposed to the mud column's hydrostatic pressure. The balance chamber therefore offsets the effect of the mud column's hydrostatic pressure. Thus, the surface closing pressure used does not have to take into account the water's depth.

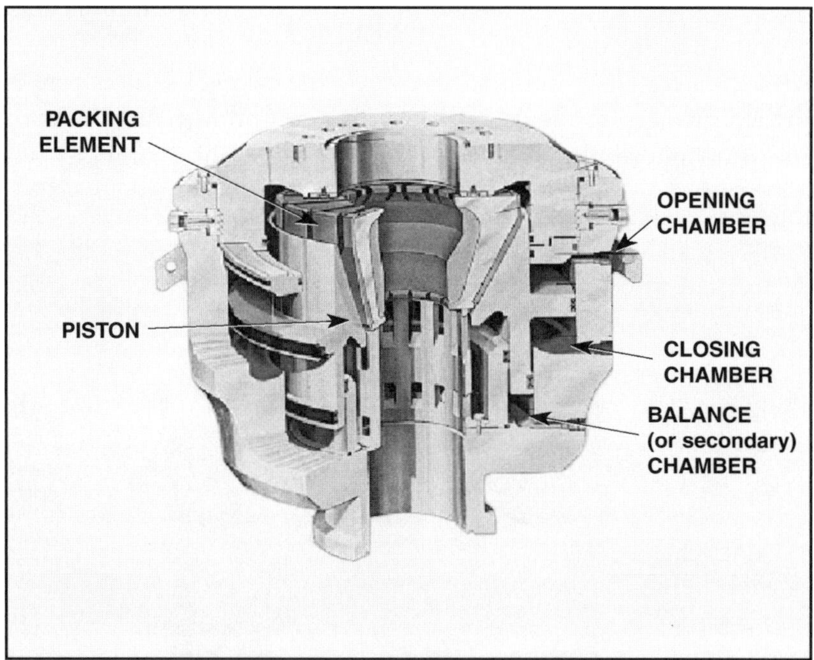

Figure 52. Hydril GL annular preventer (Courtesy of Hydril)

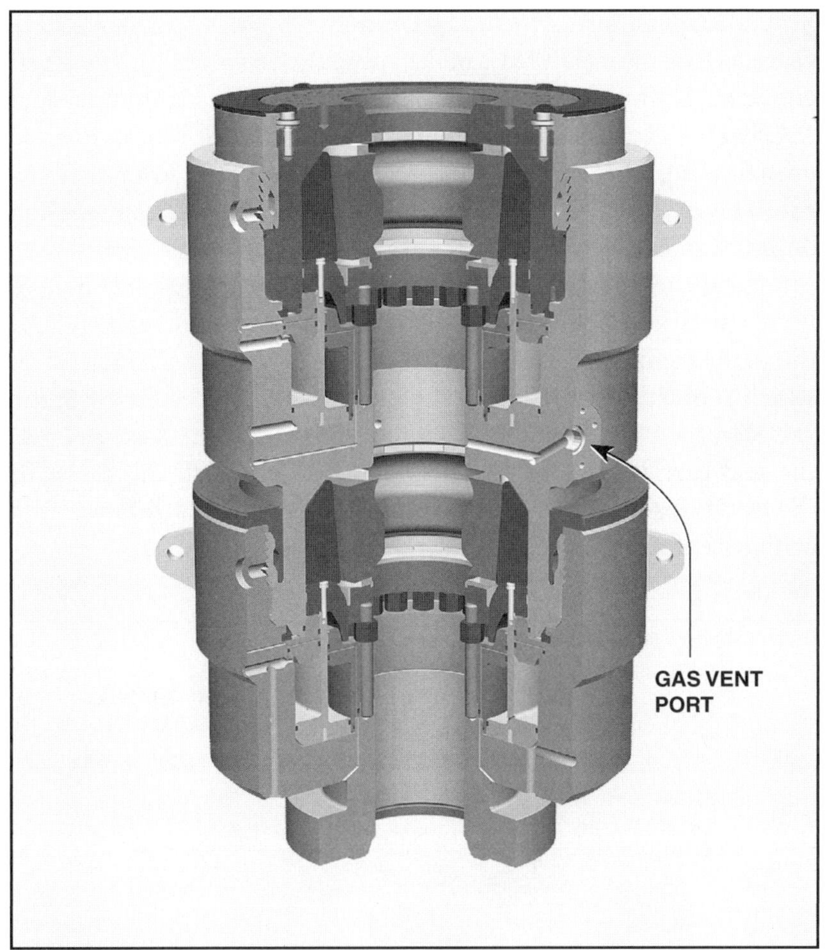

Figure 53. Cameron double unit annular BOP (Courtesy of Cameron)

Manufacturers can supply annular preventers as double units (fig. 53). Double units reduce the overall height of the assembly, because double units eliminate connections. For deepwater applications, where gas beneath a closed annular is difficult to control through the marine riser, a gas vent port is provided below the packing element (see fig. 53). The gas vent port is connected to the high-pressure choke or kill line through subsea gate valves, which allows trapped gas to be safely vented prior to opening the annular.

Subsea Gate Valves

Several manufacturers, including Cameron, Shaffer, and Worldwide Oilfield Machine (WOM), make subsea gate valves for the BOP choke and kill line side outlets. Figure 54 shows Cameron's dual MCS valve. Regardless of the manufacturer, most of these valves use a floating gate with a downstream sealing design. WOM subsea gate valves, however, use an upstream and downstream sealing design (fig. 55).

Seal carriers, which reside in machined recesses in the valve body on either side of the valve cavity, seal the valve body and the gate assembly. The gate is a small slab of steel with an orifice, which is positioned between the seal carriers to open or close the valve. In the closed position, the solid section of the gate is across the seal carrier's orifice and prevents flow through the valve. In the open position, the gate's orifice aligns with the seal carrier's orifice to allow flow through the valve.

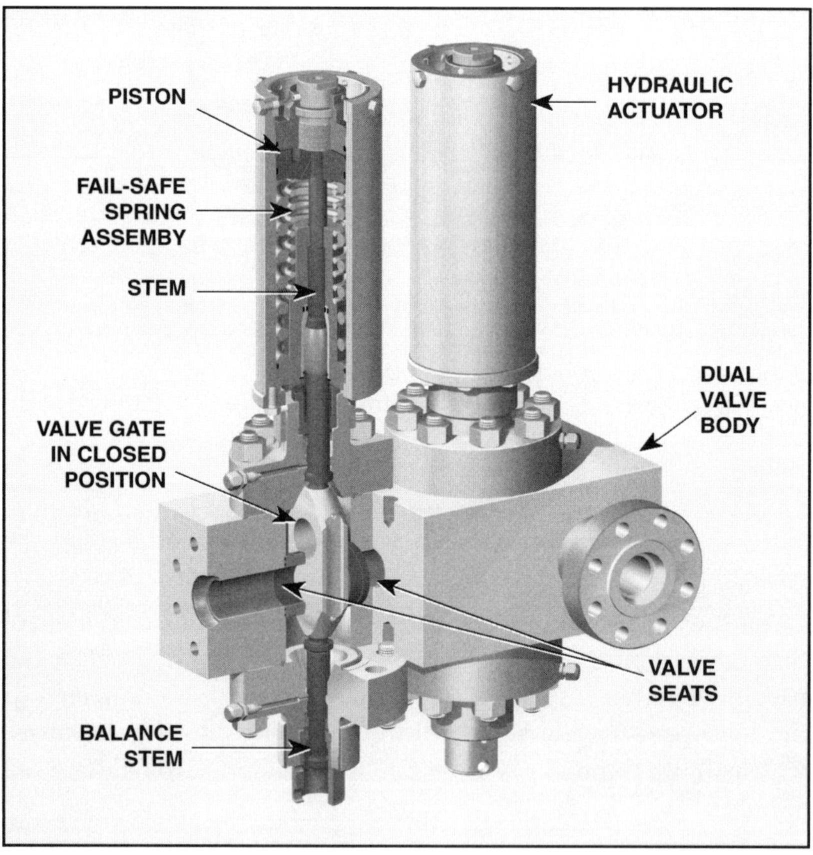

Figure 54. Cameron dual MCS valve (Courtesy of Cameron)

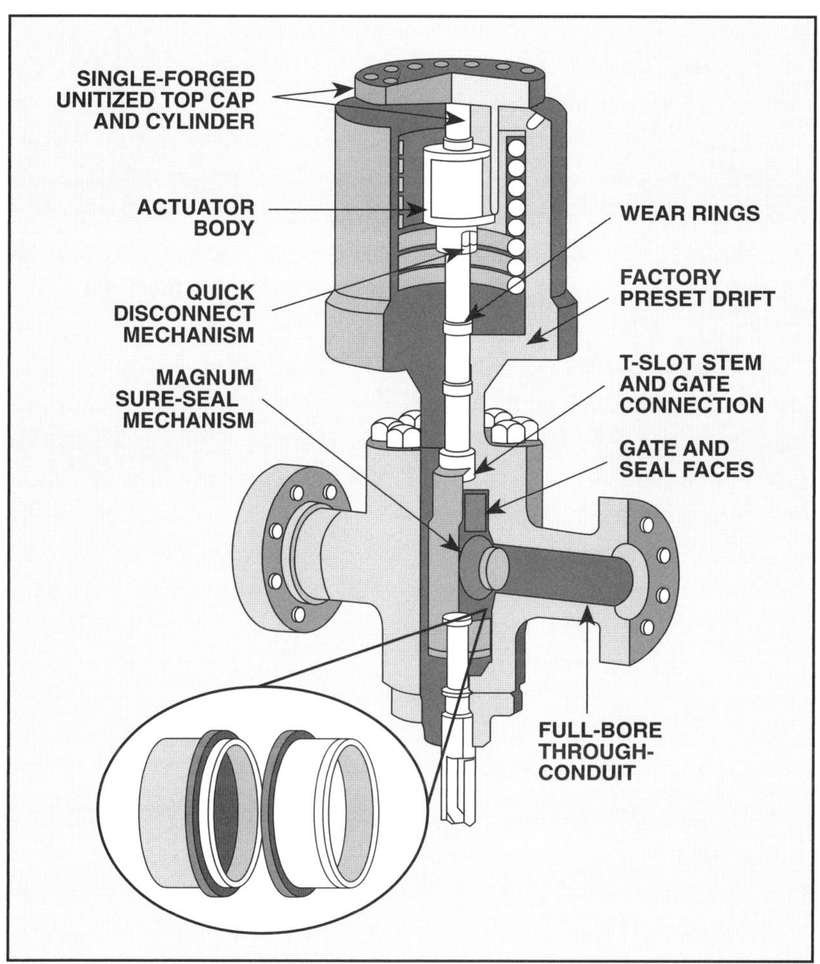

Figure 55. WOM magnum valve (Courtesy of Worldwide Oilfield Machine)

As wellbore pressure is applied to the closed valve from one direction, the pressure pushes the gate against the downstream seal carrier to create the downstream seal. When pressure is applied from the opposite direction, the gate is pushed against the other seal carrier to create a downstream seal.

Subsea choke and kill line valves are hydraulically operated by means of a hydraulic actuator attached to the body. The hydraulic actuator has a maximum working pressure of 3,000 psi (21,000 kPa). The actuator consists of a cylinder assembly, a piston, a stem that is connected to the valve gate, and a spring assembly. Since the gate is connected to the piston stem, as the piston moves, the gate moves with it. To open the valve, hydraulic pressure applied to the actuator's opening chamber moves the piston, which slides the gate across the valve cavity to open it.

To close the valve, hydraulic closing pressure applied to the actuator's closing chamber moves the piston, which slides the gate across the valve cavity to close it.

A balance stem below the gate balances the hydrostatic pressure created by seawater on the valve actuator's piston. The seawater's hydrostatic pressure must be balanced because the closing area of the actuator piston is equal to the opening area minus the stem area. Therefore, the opening area is larger than the closing area. This difference in area could allow the hydrostatic pressure of seawater to overcome the spring force and open the valve. However, the balance stem has the same area as the piston stem, and is exposed directly to the seawater's hydrostatic pressure. The balance stem therefore balances the seawater's hydrostatic pressure on the open and close sides of the piston.

The balance stem also prevents excessive pressure buildup in the valve cavity as the valve is opened. As the valve is opened, the actuator stem enters the valve cavity. This increase in metal volume in the cavity could cause a large increase in the fluid pressure, which would prevent the valve from fully opening. However, as the actuator stem enters the cavity, the balance stem exits the cavity, and the metal volume remains constant.

A large spring, housed in the actuator's chamber on the close side of the piston, provides fail-safe closure of the valve should the hydraulic system fail. Fail-safe closure is necessary to close the well and stop the flow of wellbore fluids. Note that the valve cannot fail-safe close in all situations. For example, a wellbore pressure of 500 psi (3,500 kPa) or more in the valve cavity prevents spring force from closing the valve. For this reason, a fail-safe hydraulic supply to the close side of the valve should be mounted on the BOP frame that is separate from the main control system.

To summarize—

Subsea BOP stacks—

- provide a means of pressure control when a well flow develops.
- allow drilling fluid to be circulated and conditioned.
- allow the well to be returned to a static condition.
- allow wellbore fluids to be circulated under pressure for prolonged periods of time.

- allow the drill pipe to be hung off in the BOP or at the wellhead.
- seal the well and allow wellbore pressure to be monitored.
- can shear drill pipe and allow the marine riser to be disconnected from the BOP.
- allow reconnection to the BOP and enable wellbore pressure to be monitored before opening the well.
- are rated 2K, 3K, 5K, 10K, and 15K, which stands for 2,000, 3,000, 5,000, 10,000, and 15,000 psi (13,790, 20,685, 34,475, 68,950, and 103,425 kPa).

From the bottom up, a typical subsea BOP stack consists of—

- a high-pressure wellhead connector.
- four ram preventers (in deepwater applications, up to six ram preventers may be used).
- choke and kill side outlets below each set of ram preventers.
- high-pressure valves on the choke and kill side outlets.
- high-pressure choke and kill piping.
- an annular preventer.
- a BOP mandrel.
- BOP-mounted accumulator bottles.
- a top receiver plate that houses the choke and kill line connections to the LMRP, the BOP control system receptacles, and LMRP guides.

Hydraulic connectors allow the crew to remotely connect the BOP assembly to the wellhead and the LMRP to the BOP.

Cameron hydraulic connectors include the Model 70, the HC, and the DWHC.

ABB Vetco Gray's hydraulic connector is the H4.

DrilQuip's hydraulic connector is the DX.

Hydraulic connectors consist of—

- a lower body.
- an upper body.
- a cam ring.
- locking dogs or segments.
- a seal groove.
- a hydraulic system.

Important points about hydraulic connectors include—

- replace the wellbore stainless steel gasket each time the BOP is run to the wellhead.

- ensure that the new gasket is securely retained in the connector prior to running the BOP.
- ensure that the manufacturer's recommended lubricant is used on the cam ring and the locking dogs or segments to avoid excessive friction build up.

Ram preventers—

- close around drill pipe or on open hole to seal the hole.
- are available in pressure ratings of 2,000, 3,000, 5,000, 10,000 and 15,000 psi (14,000, 21,000, 35,000, 70,000, and 105,000 kPa).

Important points to remember about ram preventers include—

- install the correct size rams for the drill pipe in use.
- record the type and size of the rams in each BOP cavity.
- accurately record the distance to each ram cavity from the rotary table.
- do not apply pressure above the ram blocks.
- do not apply opening pressure to rams when wellbore pressure is applied.
- know the capabilities of the shear rams installed.
- regularly measure the wear between the ram block and the cavity.
- regularly inspect the ram packers and top seals.
- because ram packers and top seals have various temperature ratings, use the correct ram packers and top seals for well conditions.
- do not close pipe rams or variable bore rams without pipe in the BOP; otherwise, the ram packers will be damaged.

Annular preventers—

- close and seal the wellbore and, at the same time, allow the drill stem to be moved through the closed preventer.
- are available in several pressure ratings and sizes.
- are sensitive to the water depth in which they are employed (except for Hydril GL).

Choke Manifolds

A choke manifold (fig. 56) allows the rig crew to hold back-pressure against the formation while they circulate a kick from the well. On surface choke manifold systems, crew members can usually pump only through the kill line, and only the choke line can take returns from the well. However, on choke manifold assemblies for subsea drilling applications, the manifold permits the crew to pump through or to take returns through either line. Being able to pump or take returns through either line provides redundancy in the subsea system.

A choke manifold is a series of gate valves and choke assemblies in a compact manifold that provides various flow paths for drilling fluids and well fluids encountered in well control. Normally, choke manifolds provide three flow paths: one through a remotely operated choke assembly, another through a manually operated choke assembly, and another directly to the discharge manifold or mud-gas separator. The choke manifold normally consists of a high-pressure side and a low-pressure side.

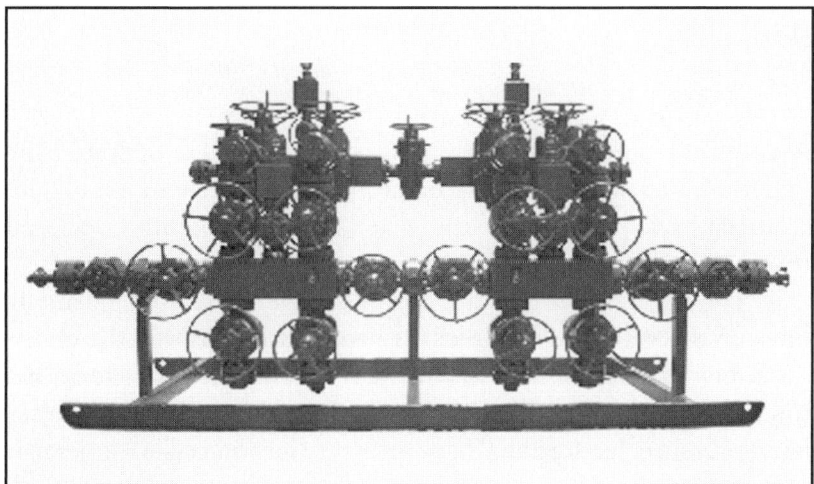

Figure 56. Choke manifold (Courtesy of Shaffer)

The high-pressure side is upstream of the chokes, and the low-pressure side is downstream of the chokes. All components on the upstream side must be constructed and installed to full-rated working pressure. However, the components on the downstream side can be of a lower rated pressure. For example, in a 10,000-psi (70,000-kPa) manifold assembly, the upstream components are rated to 10,000 psi (70,000 kPa) and the downstream components are rated to 5,000 psi (35,000 kPa). On a 15,000-psi (105,000-kPa) manifold assembly, the upstream components are rated to 15,000 psi (105,000 kPa) and the downstream components rated to 10,000 psi (70,000 kPa). A minimum of two full-rated gate valves should be installed upstream of each choke assembly and at least one gate valve downstream of each choke assembly.

The design of the choke manifold should allow for—

- circulation of the well through the drill pipe, with well returns through the kill line;
- circulation of the well through the drill pipe, with well returns through the choke line;
- circulation of the well through the drill pipe, with well returns through the choke and kill lines simultaneously;
- circulation of the well through the kill line, with well returns through the choke line; and
- circulation of the well through the choke line, with well returns through the kill line.

The gate valves, chokes, piping, flanges, tees, and elbows that make up the manifold should have a pressure rating equal to the pressure rating of the ram preventers and should have a minimum ID of 3 in. (75 mm).

Chokes

A common type of choke assembly (fig. 57) is a hydraulically operated, variable orifice valve that, when properly manipulated, controls back-pressure on the well as a kick is being circulated from the well.

When circulating a kick out of the well, the crew normally lines up the manifold to direct the return flow through the choke assembly. As return fluid flows through the choke, the choke operator, from a remote position, adjusts the choke's orifice size to alter back-pressure. If additional back-pressure is required, the operator reduces the choke's orifice size; if less back-pressure is required, the operator increases the choke's orifice size.

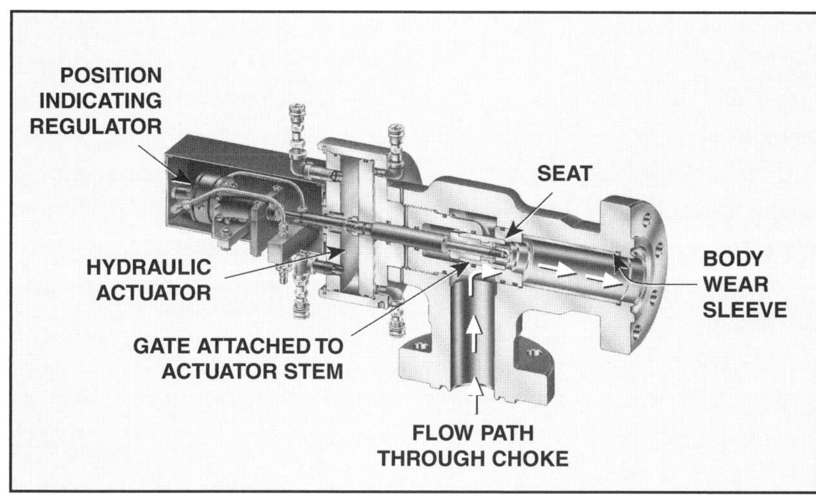

Figure 57. Hydraulically-operated choke (Courtesy of Cameron)

For safety reasons, the control panel that the operator uses to operate the hydraulic chokes (fig. 58) is mounted remotely from the manifold and close to the driller's station. Control panels include—

- a self-contained hydraulic reservoir and pump;
- control valves for operating the chokes; and
- gauges required to monitor the choke's position, drill pipe (standpipe) pressure, casing pressure, the mud pump's stroke rate, and a counter to track the number of pump strokes.

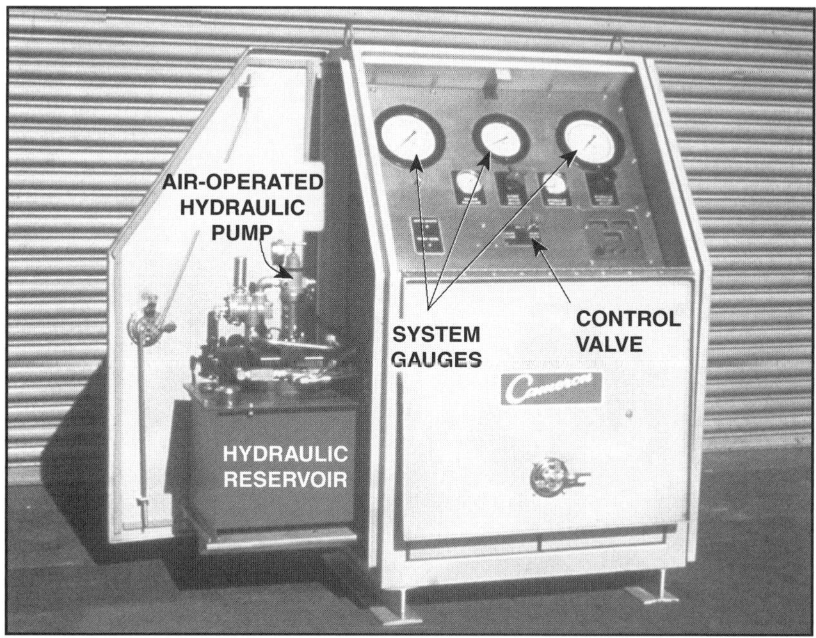

Figure 58. Choke control panel for remote operation of the hydraulic choke (Courtesy of Cameron)

The choke's hydraulic pump is air operated; so, sufficient air supply must be available to operate the pump at its maximum pressure. Further, a backup method of pump operation should be provided in the event that rig air supply fails.

Usually, a manually-operated choke (fig. 59) is provided to allow well-control operations to continue should the hydraulic remote chokes fail. The manual choke is identical to the hydraulic choke except that a crew member turns a control wheel by hand to adjust the choke's opening.

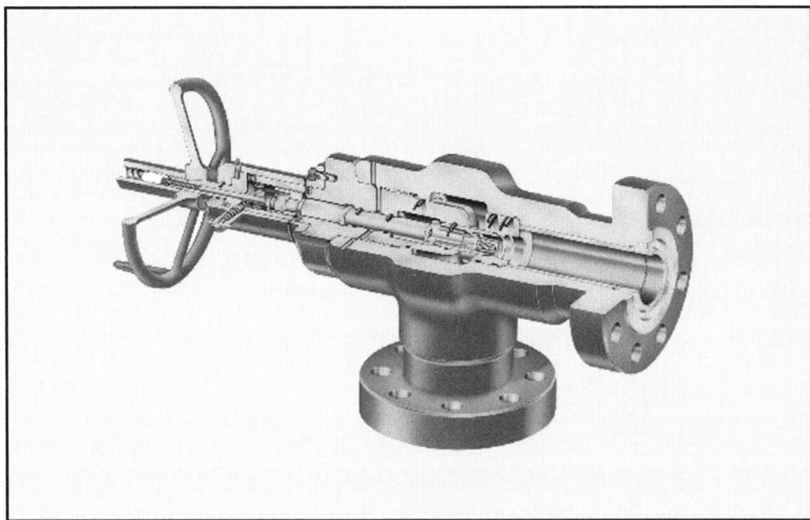

Figure 59. Manually-operated choke (Courtesy of Cameron)

Gate Valves

The gate valves in the choke manifold (fig. 60) are identical to those used on the subsea BOP choke and kill outlets, but are manually operated through a threaded, nonrising stem and hand wheel. By rotating the stem, the gate slides through the valve body to either completely open or completely close the valve. Gate valve openings are not adjustable. A nonrising stem prevents increases in valve body pressure that would occur if the stem entered the valve's body. Crew members use the valves to select the desired flow path through the manifold and to isolate failures in the piping, choke assemblies, or downstream manifold without interrupting the well's circulation.

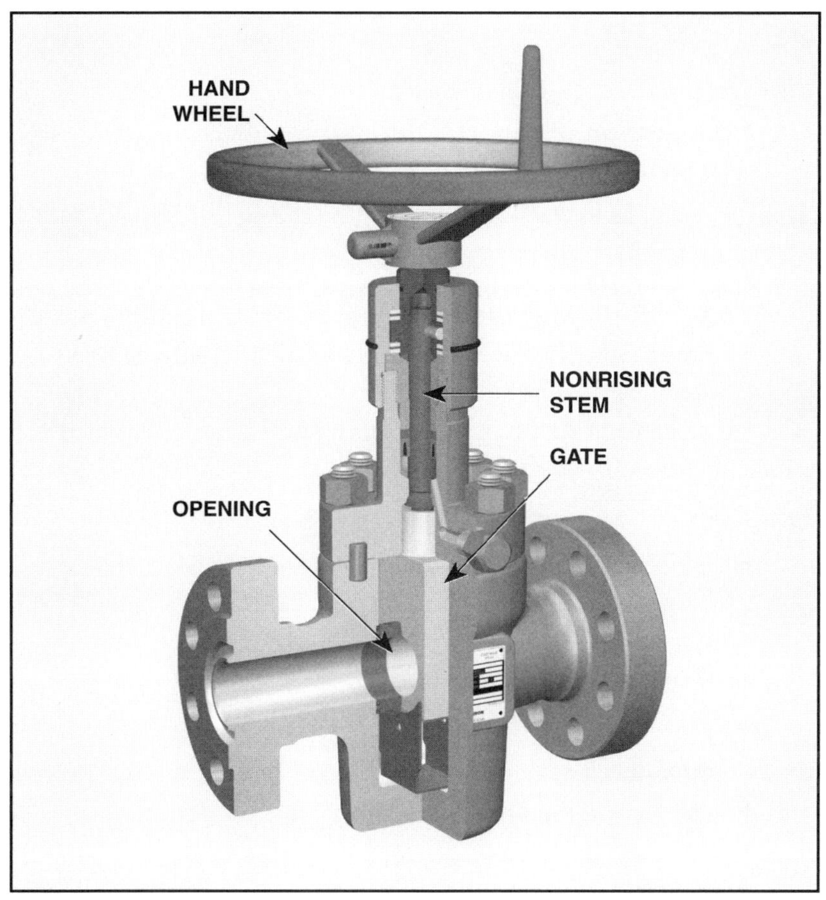

Figure 60. Gate valve used in choke manifolds (Courtesy of Cameron)

To summarize—

A choke manifold—

- allows the rig crew to hold back-pressure against the formation while they circulate a kick from the well.
- is a series of gate valves and choke assemblies in a compact manifold that provides various flow paths for drilling fluids and well fluids encountered in well control.
- provides three flow paths: one through a remotely operated choke assembly, another through a manually operated choke assembly, and another directly to the discharge manifold or mud-gas separator.
- consists of a high-pressure side and a low-pressure side.

Choke manifold design should allow for—

- circulation of the well through the drill pipe, with well returns through the kill line;
- circulation of the well through the drill pipe, with well returns through the choke line;
- circulation of the well through the drill pipe, with well returns through the choke and kill lines simultaneously;
- circulation of the well through the kill line, with well returns through the choke line; and
- circulation of the well through the choke line, with well returns through the kill line.

Chokes—

- are hydraulically operated, variable orifice valves that, when properly manipulated, control back-pressure on the well as a kick is being circulated from the well.
- allow an operator to adjust back-pressure on the well by changing the size of the choke opening.

Choke control panels include—

- a self-contained hydraulic reservoir and pump.
- control valves for operating the chokes.
- gauges required to monitor the choke's position, drill pipe (standpipe) pressure, casing pressure, the mud pump's stroke rate, and a counter to track the number of pump strokes.

Mud- and Gas-Handling Equipment

A mud-gas separator is an essential item of rig equipment to handle the mud and gas flow from the choke manifold during a gas kick. It provides a method of separating the gas from the mud and venting the gas at a safe distance from the drill floor. Once separated from the gas, the mud can be returned to the active system. Various types of separator are available, but a bottom opening to enable fluid return to the mud system and an opening to allow gas to vent from the top are essential. The gas vent line's ID should be 6 to 8 in. (150 to 200 mm) to minimize back-pressure on the separator. The vent line is normally routed up the derrick where the gas can vent safely above the crown. Usually, a bypass to an emergency overboard line is provided should the separator become overloaded or fail. Prior to returning to the active system, the mud passes through vacuum degassers to separate any small gas accumulations that were left in the mud after it passed through the mud-gas separator. Finally, the active mud tanks have large surface areas and an agitation system. The large surface areas expose the mud to the atmosphere and the agitators stir the mud. In this manner, even smaller quantities of gas that may still be left in the mud can be removed.

Since correctly weighted mud is the primary means of well control, bulk mud mixing devices, including storage tanks with a transfer system and mixing hoppers, are required to quickly raise the mud weight.

Adequate volumes of conditioned mud should be maintained in the tanks at all times, taking into consideration the possibility of lost circulation. Further, tank space should be available to allow for expansion if it becomes necessary to circulate out a gas kick.

To summarize—

A mud-gas separator—

- handles the mud and gas flow from the choke manifold during a gas kick.
- separates the gas from the mud and vents the gas at a safe distance from the drill floor.
- returns mud to the active system through a bottom opening.
- has an opening at the top to vent gas.
- usually vents gas to the crown.

Auxiliary Surface Well-Control Equipment

Auxiliary surface well-control equipment includes equipment for the drill string and various surface devices. Equipment needed for the drill string include drill pipe safety valves, upper and lower kelly cocks, inside BOP valves, and drill pipe float valves. Additional surface equipment includes a trip tank, mud pit volume measuring and recording devices, a flow-line mounted flow-line sensor, and safety valves for the top drive.

Drill Pipe Safety Valve

A full-opening drill pipe safety valve with the same pressure rating as the ram preventers should be available on the rig floor at all times, complete with its closing wrench. The required crossover subs should also be available to allow the valve to be connected to any section of the drill string, casing, or tubing in use. The drill pipe safety valve's OD should be of a size that allows it to be run in the hole. Furthermore, it should have a top connection that allows it to be connected to the top drive or kelly.

Inside Blowout Preventers

An inside blowout preventer is a drop-in check valve that the floor crew installs on the surface. It is also called a Gray valve after one of the manufacturers of such valves. It should be available on the drill floor at all times. If the drop-in type is used, a landing sub is positioned in the drill string at the drill collars. A dart of the correct size to match the landing sub is kept on the rig floor at all times. Note that the dart must pass through all components of the drill string above the landing sub.

Upper and Lower Kelly Cocks

The upper and lower kelly cocks should have a pressure rating equal to the ram preventers. The upper kelly cock is installed between the swivel and the kelly, and, when closed, isolates the swivel's packing, the rotary hose, and the standpipe from pressure in the drill string. The lower kelly cock is installed immediately below the kelly and can be closed to isolate the kelly.

Float Valves

A float is placed in the drill string to prevent upward flow of fluid inside the drill string. The float valve is a special type of back-pressure, or check, valve. It is usually placed in the drill string between two drill collars or between the drill bit and the drill collars. Since the float valve prevents the drill string from being filled with fluid through the bit as it is run into the hole, the drill string must be filled from the top to prevent collapse of the drill pipe.

Types of float valves include—

- a flapper-type float valve, whose ID is about the same as the tool joint. Consequently, a flapper-type valve permits the passage of balls and go-devils, which may be required for operation of tools inside the drill string below the float valve.
- a spring-loaded ball- or dart-and-seat float valve. Although its construction does not allow tools to pass, a ball- or dart-and-seat float valve provides instantaneous and positive shutoff of backflow through the drill string.

Neither valve is full bore and thus cannot sustain long duration or high-volume pumping of drilling fluid or kill fluid. However, a wireline-retrievable valve is available, which seals in a profiled body and whose ID is the same as a tool joint. This valve may be used to provide full access, if required.

Standpipe Manifolds

The standpipe manifold is a part of the mud circulation system and consists of several valves, piping, and fittings such as tees and elbows. The standpipe conducts mud from the pumps to the kelly or top-drive swivel. Normally, rigs use two standpipes in the derrick. Usually, one is used as a backup in case one fails. A flexible rotary hose connects the standpipe to the kelly or top-drive swivel.

The manifold valves direct the mud from the pumps to the appropriate standpipe, and isolate the mud pumps from the system when required. The standpipe manifold's pressure rating is less than the ram preventers, normally 5,000 or 7,500 psi (35,000 or 52,500 kPa). Crew members can isolate the standpipe from excessive wellbore pressure by closing the kelly cock or drill string safety valve. A connection and line to the choke manifold is provided to allow mud to be circulated from the mud pumps through the choke or kill lines. A check valve in the line prevents excessive choke manifold pressure from entering the standpipe manifold.

Trip Tanks

A trip tank is a low-volume, calibrated tank. Most trip tanks hold fewer than 100 barrels (bbl) or 15 cubic metres (m^3) of fluid. They are often calibrated in increments of ½ bbl or in L. A trip tank can be isolated from the surface mud system and accurately monitor the amount of mud going into or returning from the well. A trip tank may be of any shape, provided that crew members can read the volume contained in the tank at any point. The readout may be at the tank or from a remote position such as the driller's station: preferably, both will be available. The size and configuration of the tank should be such that volume changes of ½ bbl (L) can be easily detected by the readout arrangement. Tanks containing two compartments with monitoring arrangements in each compartment are preferred because such an arrangement facilitates removing or adding drilling fluid without interrupting rig operations.

As the name suggests, the trip tank accurately monitors mud levels in the well while tripping in or out of the hole. A trip tank can also be used to measure mud or water volume put into the well when returns are lost. It can also monitor hole volumes while logging or following a cement job. A trip tank can also be used to calibrate the amount of mud a pump puts out for each stroke. While a trip tank can be used to measure the volume of mud bled from or pumped into the well as pipe is stripped into or out of the well, a separate and smaller stripping tank is normally used for this purpose.

Mud Pit Volume Measuring Devices

Automatic mud pit volume measuring sensors transmit either a pneumatic or electrical signal from the mud pits to recorders and alarm devices at the driller's station. These devices monitor pit levels and detect fluid gain or loss. The pit level alarm can be adjusted to sound when the pit level changes by a predetermined amount.

Flow Rate Sensors

A flow rate sensor is mounted in the mud return flow line. It monitors the rate at which the mud returns from the well. Because an increase in return flow can indicate a well kick, it gives early detection of formation fluids entering the wellbore. Similarly, because a decrease in return flow can indicate lost returns, a flow rate sensor can warn of lost circulation.

Top-Drive Safety Valves

Two safety valves on the top drive serve the same purpose as the kelly cocks. The upper safety valve is pneumatically or hydraulically operated and controlled at the driller's console. The lower valve is a standard kelly cock and is manually operated.

Generally, if it becomes necessary to prevent or stop flow up the drill pipe during tripping operations, a separate drill pipe valve should be used rather than either of the top-drive valves. However, if flow up the drill pipe prevents stabbing a separate drill pipe valve, the top-drive valves can be used.

To summarize—

Auxiliary equipment for the drill string includes—

- drill pipe safety valves.
- upper and lower kelly cocks.
- inside BOP valves.
- drill pipe float valves.

Additional auxiliary surface equipment includes—

- a trip tank.
- mud pit volume measuring and recording devices.
- a flow-line mounted flow-line sensor.
- top-drive safety valves.

Drill pipe safety valves should—

- have the same pressure rating as the ram preventers.
- be available on the rig floor at all times, complete with closing wrench.
- have crossover subs available to allow the valve to be connected to any section of the drill string, casing, or tubing in use.
- have a top connection that allows them to be connected to the top drive or kelly.

An inside blowout preventer—

- is a drop-in check valve that the floor crew installs on the surface.
- should be available on the drill floor at all times.

Upper and lower kelly cocks—

- should have a pressure rating equal to the ram preventers.
- close off the drill stem above and below the kelly to prevent pressure from entering the swivel.

A float valve—

- is placed in the drill string to prevent upward flow of fluid inside the drill string.
- may be either a flapper type or a spring-loaded dart valve.

Standpipe manifolds—

- are part of the mud circulation system and consist of several valves, piping, and fittings such as tees and elbows.
- conduct mud from the pumps to the kelly or top-drive swivel.
- should have a pressure rating of 5,000 or 7,500 psi (35,000 or 52,500 kPa).

Trip tanks—

- are low-volume, calibrated tanks.
- accurately monitor the amount of mud going into or returning from the well.

Mud pit volume measuring devices—

- transmit either a pneumatic or electrical signal from the mud pits to recorders and alarm devices at the driller's station.
- monitor pit levels and detect fluid gain or loss.

Flow rate sensors—

- are mounted in the mud return flow line.
- monitor the rate at which the mud returns from the well.
- give early detection of formation fluids entering the wellbore.

Top-drive safety valves—

- serve the same purpose as the kelly cocks.

Hydraulic Control Systems

Every component in the subsea BOP assembly is hydraulically operated and a control system is used to operate them. The BOP control system must deliver, on command, hydraulic fluid at the correct pressure to operate the BOP components. Further, when a component is operated, the system must also vent hydraulic fluid from the opposite side of the operated component. Subsea BOP control systems are indirect pilot-operated systems. That is, hydraulic pilot signals from the surface operate various valves and regulators mounted in the subsea control pods that then supply the operating fluid to the BOP components. Two types of indirect pilot system are the straight hydraulic type and the multiplexed type. The straight hydraulic system (fig. 61) is rated to 3,000 psi (21,000 kPa) working pressure and delivers the hydraulic pilot signals from the surface through pilot hoses to the valves and regulators mounted in the subsea control pods.

The multiplexed system is usually rated to 5,000 psi (35,000 kPa) working pressure and sends electronic signals from the surface to subsea-mounted solenoid valves that then send the pilot signals to the valves and regulators.

Straight Hydraulic Control Systems

BOP control system manufacturers include Koomey, Shaffer, Hydril, ABB Systems, and Cameron. Basically, all straight hydraulic systems operate the same way, although different manufacturers may use slightly different valves and regulators to achieve the same result.

The control system can be separated into four individual sections—

1. the surface accumulator unit, or HPU;
2. the hose bundles, hose reels, and hose reel control panels;
3. the subsea control pods; and
4. the remote electric panels.

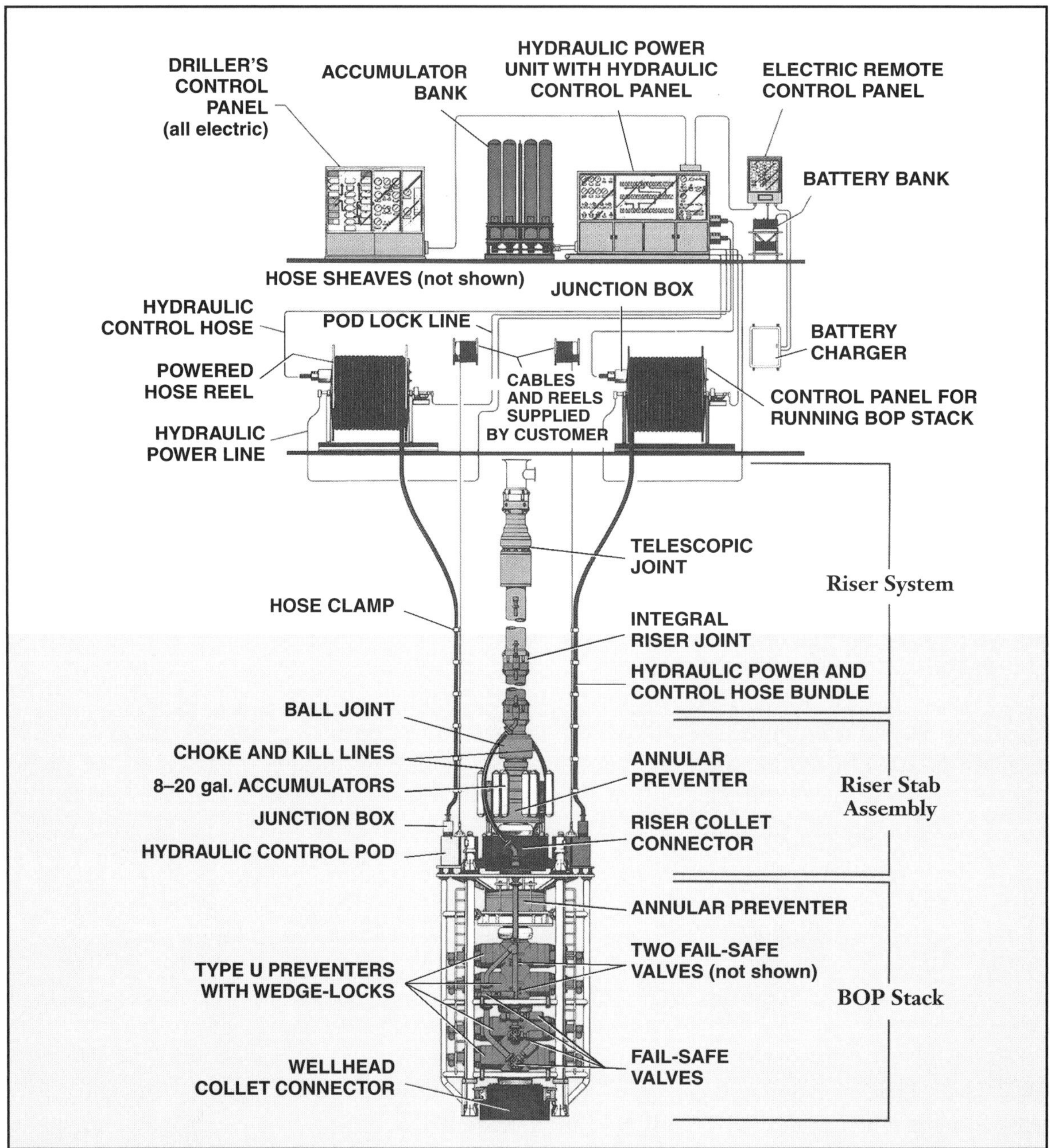

Figure 61. Hydraulic control for subsea BOPs (Courtesy of Cameron)

Surface Accumulator Units

The surface accumulator unit (figs. 62, 63) contain hydraulic fluid reservoirs, a fluid mixing system, pumps, accumulator storage bottles, a flowmeter, a pod-select valve, a pilot manifold, hydraulic junction boxes, pressure gauges, and electrical junction boxes required to operate the BOP components.

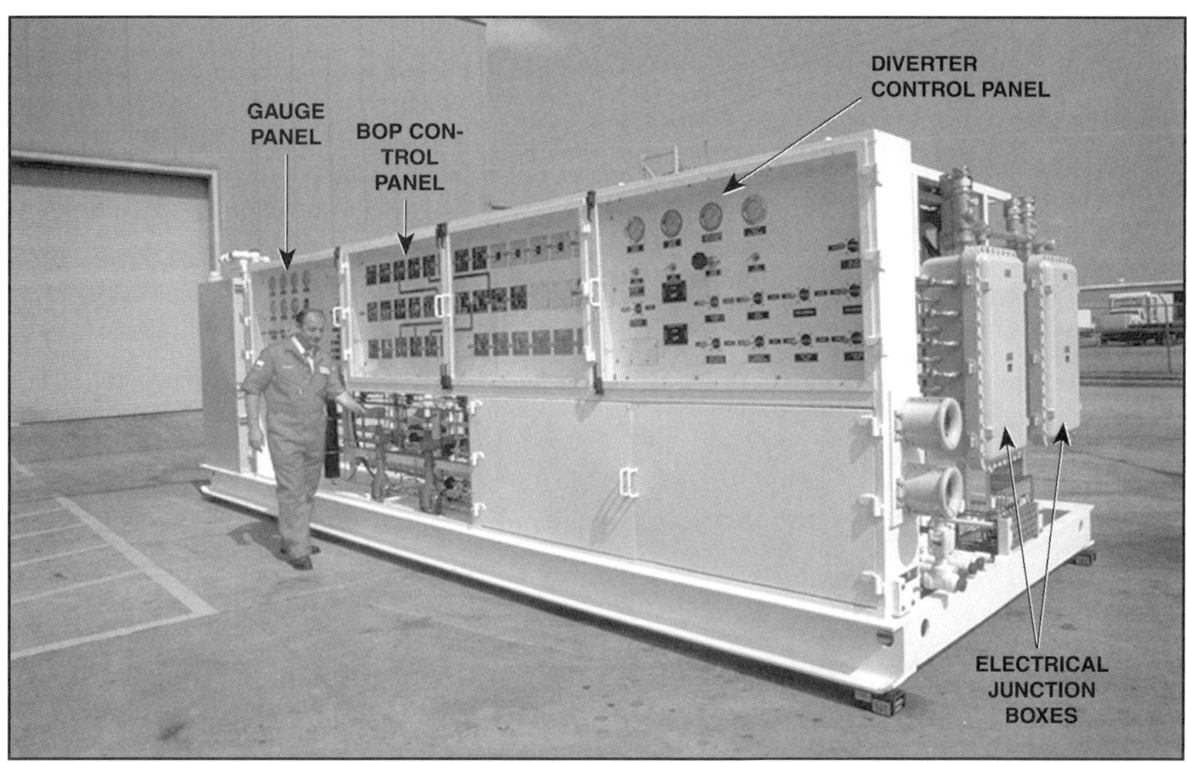

Figure 62. Surface accumulator unit (front view) (Courtesy of Cameron)

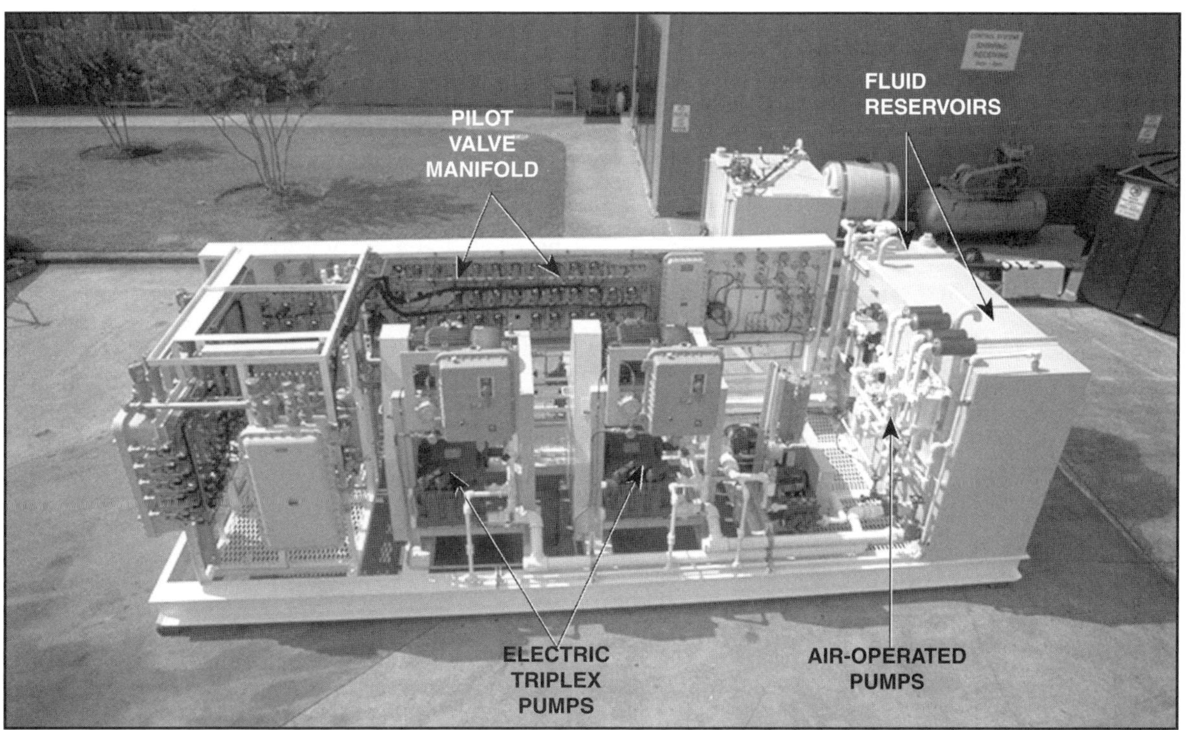

Figure 63. Surface accumulator unit (back view) (Courtesy of Cameron)

Fluid Reservoirs and Fluid Mixing System

Two fluid reservoirs are mounted on the accumulator unit's skid. One reservoir stores the fluid concentrate and the other stores the mixed fluid. The mixed fluid reservoir has a storage capacity that is twice the usable fluid capacity of the accumulators.

BOP operating fluids are water-based concentrates that are mixed with fresh water to provide the correct level of corrosion protection, lubricity, and protection against bacteria.

A fluid mixing system is mounted at the reservoirs and is operated by a float level switch in the mixed fluid reservoir. When the fluid level in this reservoir drops to a certain level, the float drops to start up the mixing system. The mixing system allows fresh water to flow into the reservoir and draws concentrate from the concentrate reservoir into the fresh water at the correct ratio. When the mixed fluid reservoir is topped up, a second float shuts down the mixing system. A third reservoir can be provided for monoethylene glycol, which may be required to provide freeze protection in cold climates.

The mixed fluid reservoir is connected to the pumping system through large-diameter low-pressure piping, valves, and suction strainers. It is vitally important to the efficient operation of the entire system that the fluid is mixed to the correct ratio and is strained and filtered.

Pumping System

Two separate pumping systems are provided. The primary pumps are electrically driven triplex pumps and the secondary pumps are usually pneumatically driven. The combination of primary and secondary pumps must have sufficient output to pressurize the entire system from zero to maximum working pressure in 15 minutes or less. A pressure switch controls the primary electric pumps. The switch starts the pumps when maximum system pressure falls to about 10 percent below maximum system pressure. When maximum pressure is reached, the switch shuts them down. Relief valves protect the pumps and piping from excess pressure should the pressure switch fail to shut the pumps down.

In some instances, both the primary and secondary pumps are electrically operated. Separate distribution panels send power to each pump. The pumps deliver fluid through high-pressure filters to banks of accumulator storage bottles. The bottles store a sufficient amount of fluid to operate the BOP functions with the pumps out of service. Such storage ensures that the BOP can be operated in the event the pumping system is lost.

Accumulator Bottles

Accumulator bottles (fig. 64) are mounted in individual banks of up to 24 in each bank. Each bank has separate isolation and bleed valves. Having separate valves allows repairs to be carried out without having to shut down the entire system. The number of accumulator bottles in the system depends on the total fluid volume required to operate the BOP components.

Figure 64. Accumulator bottles (Courtesy of Cameron)

Accumulators use either a bladder or a float to separate the operating fluid from a nitrogen precharge. Bottle capacity ranges from 1 to 35 gallons (gal) or 4 to 140 L. In general, 15-gal (60-L) accumulators are used for fluid storage on the accumulator unit. Each accumulator bottle is precharged with nitrogen that forces the fluid out of the accumulator at very high rates. Nitrogen precharge pressure is 1,000 psi (7,000 kPa) for 3,000-psi (21,000-kPa) working pressure systems and 1,500 psi (10,500 kPa) for 5,000-psi (35,000-kPa) working systems. In a bladder-type accumulator, a rubber bladder, which is charged with nitrogen, separates the nitrogen from the fluid. As the pumps increase the fluid pressure in the accumulator, the bladder is compressed until nitrogen precharge pressure equals fluid pressure. When fluid exits the accumulator to a BOP component, the bladder expands to force the fluid out of the accumulator. The float-type accumulator has a float arrangement that separates the nitrogen precharge from the fluid.

Flowmeter

A fluid flowmeter measures the amount of fluid exiting the accumulator bottles. It is mounted in the line that connects the accumulator banks to the pod-select valve. By mounting the flowmeter in this position, it can precisely record the amount of fluid being delivered to the subsea control pods. The flowmeter is one tool to use for troubleshooting subsea BOP malfunctions.

For example, if a ram preventer is to be closed, the flowmeter can be set at 0 gal (0 L) and the ram closed. The flowmeter records the amount of fluid delivered, and if this amount is less than the amount known to be required for closing the rams, the operator can tell immediately that the rams have not fully closed. On the other hand, if the flowmeter records that more fluid is being pumped than is required to close the rams, then the operator can assume that a leak exists in the ram closing circuit.

Pod-Select Valve

Because two identical subsea control pods, which are designated yellow and blue for identification, provide redundancy to the system, a pod-select valve is required to distribute the 3,000-psi (21,000-kPa) accumulator fluid from the surface to the selected subsea control pod. All the surface accumulator banks are connected to a common line that supplies a special valve mounted at the accumulator unit's front panel. This valve is a 1-in. (25.4-mm) manipulator-type valve. It has three positions and is a four-way valve. When set to its center position, the valve blocks accumulator pressure and vents both control pods. This position allows personnel to work on the control pods or can isolate a subsea fluid leak.

The pod-select valve is also ported so that when the yellow pod is selected and receiving accumulator pressure, the blue pod is vented. Similarly, when the blue pod is selected and receiving accumulator pressure, the yellow pod is vented. This operation ensures that only one subsea control pod can receive accumulator pressure. The pod-select valve can be operated manually by using the valve handle to position the valve, or remotely by a pneumatic cylinder connected to the valve handle.

Pilot Manifold

A regulator is installed in the supply line from the accumulator banks to the pod-select valves. It is placed in the line before the flowmeter and a pilot manifold pipe is connected to the regulator's outlet port. It reduces (regulates) 3,000-psi (21,000-kPa) accumulator pressure to 2,000 psi (14,000 kPa) and sends it to a pilot manifold (fig. 65). The pilot manifold consists of several ¼-in. (6.35-mm) pilot control valves (fig. 66). These valves have three positions and are four-way manipulator-type valves. They are mounted on the accumulator unit's front panel and connect to pilot regulators, which are usually mounted below the valves (see fig. 65).

Pilot valves operate the same way as the pod-select valve. They send the pilot signals to valves on the subsea control pods. Pilot valves can be operated manually with a handle or pneumatic actuators can operate them remotely. The pilot signals are sent to the subsea control pod valves at a regulated pressure of 2,000 psi (14,000 kPa) to provide the fastest possible signal time.

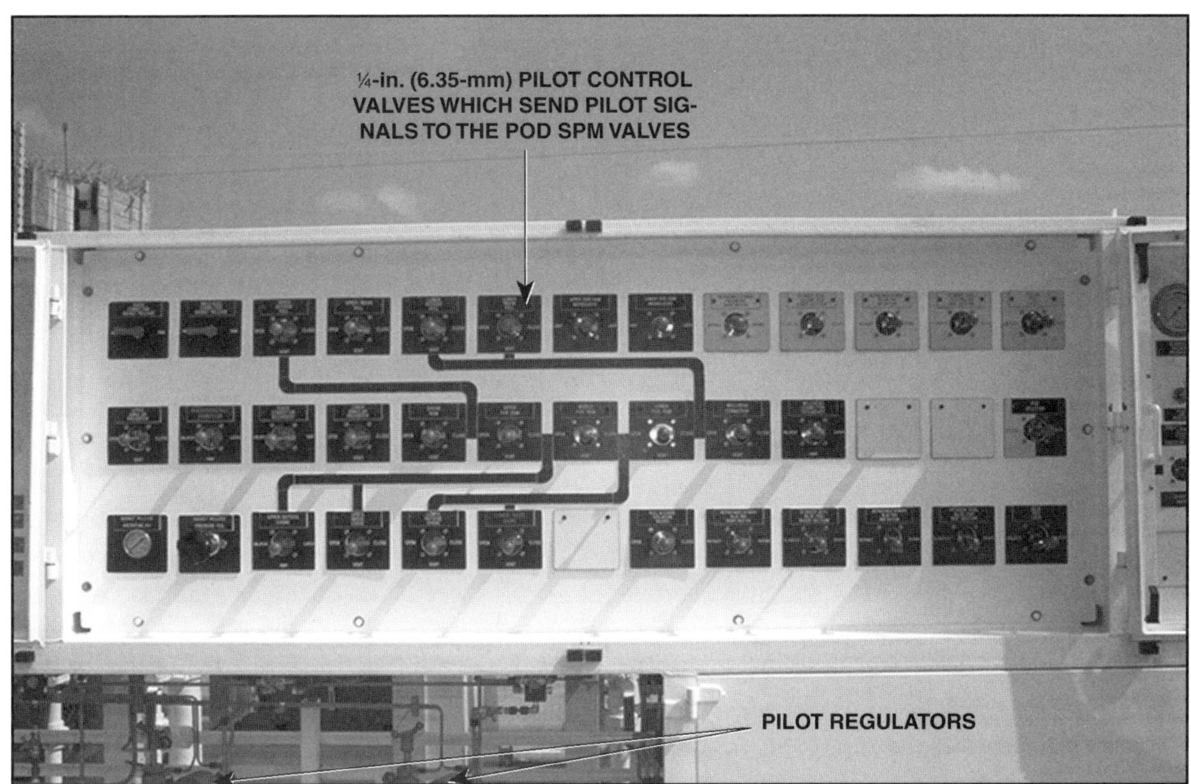

Figure 65. Pilot manifold (Courtesy of Cameron)

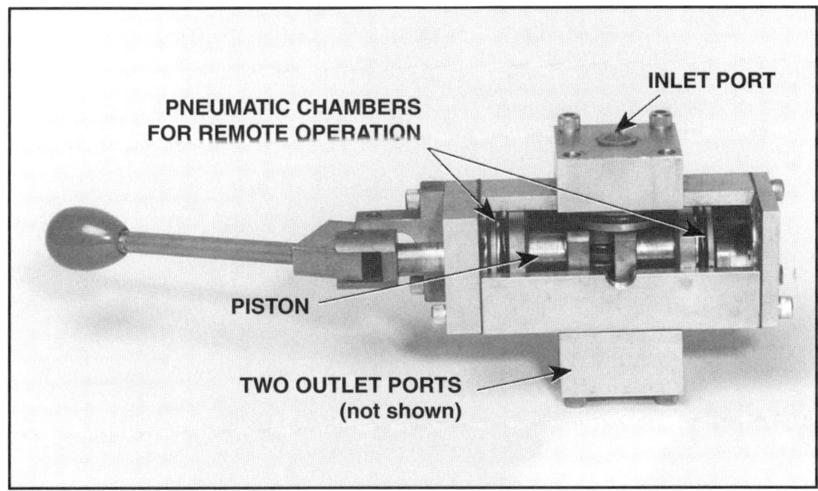

Figure 66. Pilot valve (Courtesy of Cameron)

Pilot regulators send variable pressure pilot signals, from 0 to 3,000 psi (21,000 kPa), to regulators mounted on the subsea control pod. The pilot regulators are pneumatic pilot operated and have a 30 to 1 ratio, which means that the hydraulic output pressure is 30 times the air pilot signal. For example, if an air pilot signal of 50 psi (350 kPa) is applied, the hydraulic output pressure will be 1,500 psi (10,500 kPa). The pilot accumulators provide pilot fluid storage capacity for the pilot system and normally consist of two 5-gal (20-L) accumulators.

Hydraulic Junction Boxes

High-pressure piping that has an ID of 1 in. (25.4 mm) delivers accumulator fluid from the pod-select valve's outlet ports to two hydraulic junction boxes mounted on the accumulator unit. One junction box is for the yellow pod supply and the other is for the blue pod supply. High-pressure ¼-in. (6.35-mm) piping delivers the pilot signals from the ¼-in. (6.35-mm) pilot valve outlet ports to the hydraulic junction boxes. Pilot signals are simultaneously delivered to both the yellow and the blue junction boxes. High-pressure ¼-in. (6.35-mm) piping also simultaneously delivers the variable pilot signals from the pilot regulator outlet ports to both hydraulic junction boxes.

The hydraulic junction boxes consist of two circular plates that are bolted together. The female plate contains the female section of a quick-disconnect coupling, and the male plate contains the male section of the quick-disconnect coupling. All the high pressure piping from the pod-select valve, pilot valves, and regulators is connected to the male quick-disconnect couplings. Individual hoses within the hose umbilical are connected to the female quick-disconnect coupling. An alignment pin between the two plates ensures that the plates connect properly.

Pressure Gauges

A series of pressure gauges on the front panel of the accumulator unit (fig. 67) monitor system pressures, such as accumulator pressure, pilot pressure, BOP manifold pilot pressure, BOP read-back pressure, and so forth. These gauges should be calibrated regularly to ensure accuracy.

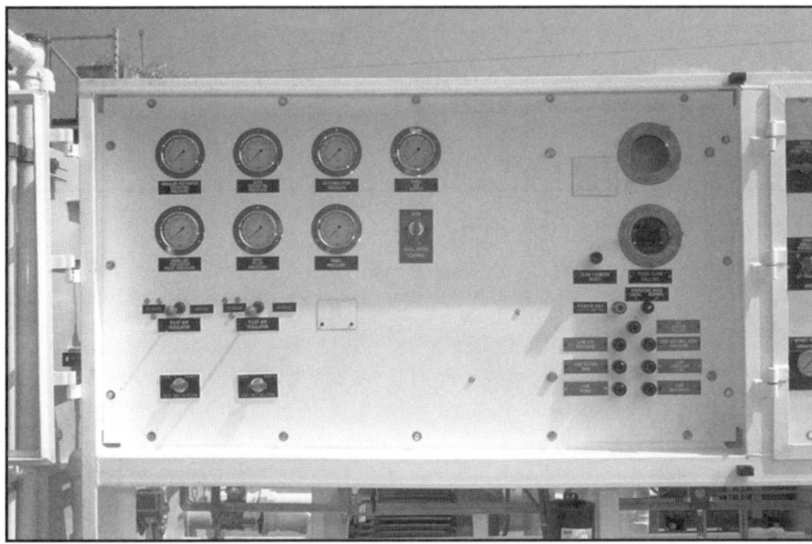

Figure 67. Pressure gauges (Courtesy of Cameron)

Electrical Junction Boxes

Explosion-proof electrical junction boxes on the accumulator unit provide the interface for the remote electrical panels and the accumulator unit's pod-select valve, pilot valves, and various system pressure readouts. Normally, two electrical junction boxes are present. One houses air-actuated solenoids that, when energized, send an air signal to the pod-select valve and pilot valves to allow remote operation. The other houses pressure switches that illuminate the indicating lights on the remote panels. This junction box also houses pressure transducers that convert variable hydraulic pressure into variable voltage output to provide a pressure readout on the remote panel's pressure gauges.

Hose Bundles, Hose Reels, and Reel Control Panels

The fluid delivered to the hydraulic junction boxes on the accumulator unit passes through quick-disconnect couplings on the junction boxes and into individual hoses in a hose bundle. The hose bundle carries hydraulic fluid to hydraulic junction boxes mounted on hose storage reels. The hose bundle is made up of a 1-in. (25.4-mm) accumulator pressure hose in the center; several 3⁄16-in. (4.76-mm) pilot hoses surround the accumulator pressure hose. An outer polyurethane sheath protects the individual hoses from handling or other damage.

The number of pilot hoses in the bundle depends on the number of hydraulically operated components on the BOP. Hose bundles are available in lengths of 5,000 ft (1,500 m) and longer, and have a pressure rating of 3,000 or 5,000 psi (21,000 or 35,000 kPa).

The hose bundles between the accumulator unit and the hose reels are generally known as jumper hoses. One end of the subsea section of hose bundle that is stored on the hose reels is connected to the jumper hose through a hydraulic junction box mounted on the side of the reel. The other end is connected to a junction plate at the subsea control pod.

The hose reels store the subsea section of hose bundle and pay out or reel in the hose as the BOP is being run or retrieved. Two reels are used; one for the yellow control pod and the other for the blue control pod. Pneumatic motors drive the reels and each reel has a braking system and a mechanical lock pin. The reels can store 5,000 ft (1,500 m) of hose and can be provided with a level wind to spool the hose neatly on the drum. Neat spooling protects the hose from damage. Large radius hose sheaves on the subsea section of the hose bundle prevent the hoses from being bent and damaged. Hose bundle clamps secure the hose to the control pod's retrieving line, or to the marine riser, as the BOP is being run. The clamps prevent damage caused by rough seas, waves, and currents.

The female section of the hydraulic junction box is mounted on the side of the reel to which the subsea section of hose bundle is connected. The jumper hose from the accumulator unit is connected to the male section. Such an arrangement allows the jumper hose to be quickly and easily removed or connected prior to running or retrieving the BOP.

The jumper hoses have to be disconnected from the hose reels as the BOP is being run or retrieved; however, accumulator supply and pilot signals to the control pods must be maintained.

For example, pressure to components such as the LMRP connector lock and the wellhead connector unlock and lock must be supplied to operate them. So, reel-mounted control panels are provided. On the control panels are ¼-in. (6.35-mm), three position, four-way selector pilot valves and a manual regulator. The 1-in. (25.4-mm) accumulator supply hose in the jumper hose bundle does not terminate at the hose reel junction box. Instead, it is connected to a high-pressure swivel union mounted on the reel's center shaft. An internal fluid passage in the reel's center shaft carries the accumulator supply into the reel, where it is terminated to the 1-in. (25.4-mm) accumulator supply hose in the reel's hose bundle. This arrangement allows accumulator supply to be maintained to the control pod, even with the reel's junction box disconnected. A tee from the reel's center shaft is connected to the reel's control panel manifold pipe. The reel's manifold pipe then supplies 3,000-psi (35,000-kPa) accumulator pressure to the manual regulator and ¼-in. (6.35-mm) pilot valves, allowing control of selected BOP components.

BOP Control Pods

Two identical BOP hydraulic control pods are located on the LMRP's base plate. They are designated as yellow and blue. The hose bundle terminates in a junction plate on the top of each pod. Stainless steel tubing or high-pressure hoses carry the fluid from the junction plate to the pod valves and regulators.

Normally, control pods can be retrieved individually to allow repairs to be carried out without retrieving the LMRP. The retrievable (male) portion of the pod contains the regulators and SPM valves required to operate the BOP components and the sealing mechanism that is required between the retrievable section and the nonretrievable receptacles. Two nonretrievable female receptacles are employed: one is the LMRP receptacle, which is bolted to the LMRP's base plate. The other is the BOP receptacle, which is bolted to the BOP's upper receiver plate. An LMRP stinger on the male section of the pod mates with the LMRP's receptacle, and a BOP stinger mates with the BOP's receptacle. This stinger-and-receptacle arrangement allows for retrieval of the male pod section and for the separation of the LMRP from the BOP. The LMRP stinger and receptacle are ported to direct fluid from the pod's SPM valves to LMRP-mounted components such as the LMRP connector and the annular preventer. The BOP stinger and receptacle are ported to direct fluid from the pod's SPM valves to the BOP's mounted components.

The BOP and LMRP stingers are extended or retracted by hydraulic cylinders. In the extended position (fig. 68), the stingers are stabbed into receptacles (fig. 69). Once seated in the receptacle, full hydraulic communication between the pod and the BOP and LMRP components occurs. To disconnect the LMRP from the BOP, the BOP stinger is hydraulically retracted into the pod from the receptacle prior to disconnecting. Retracting the stinger protects it from damage. To retrieve a pod for repair, both stingers are retracted into the pod from the receptacles. Retracted stingers are housed inside test receptacles located in the pod. These test receptacles protect the stingers and allow the pod to be tested.

The stinger's ports (fig. 70) must be sealed to the receptacle ports to allow full fluid flow to the components. Therefore, each port of the stinger has a packer seal that a threaded seal retainer keeps in the port. These seals are hydraulically energized when the stinger is located in the receptacle, and hydraulically deenergized

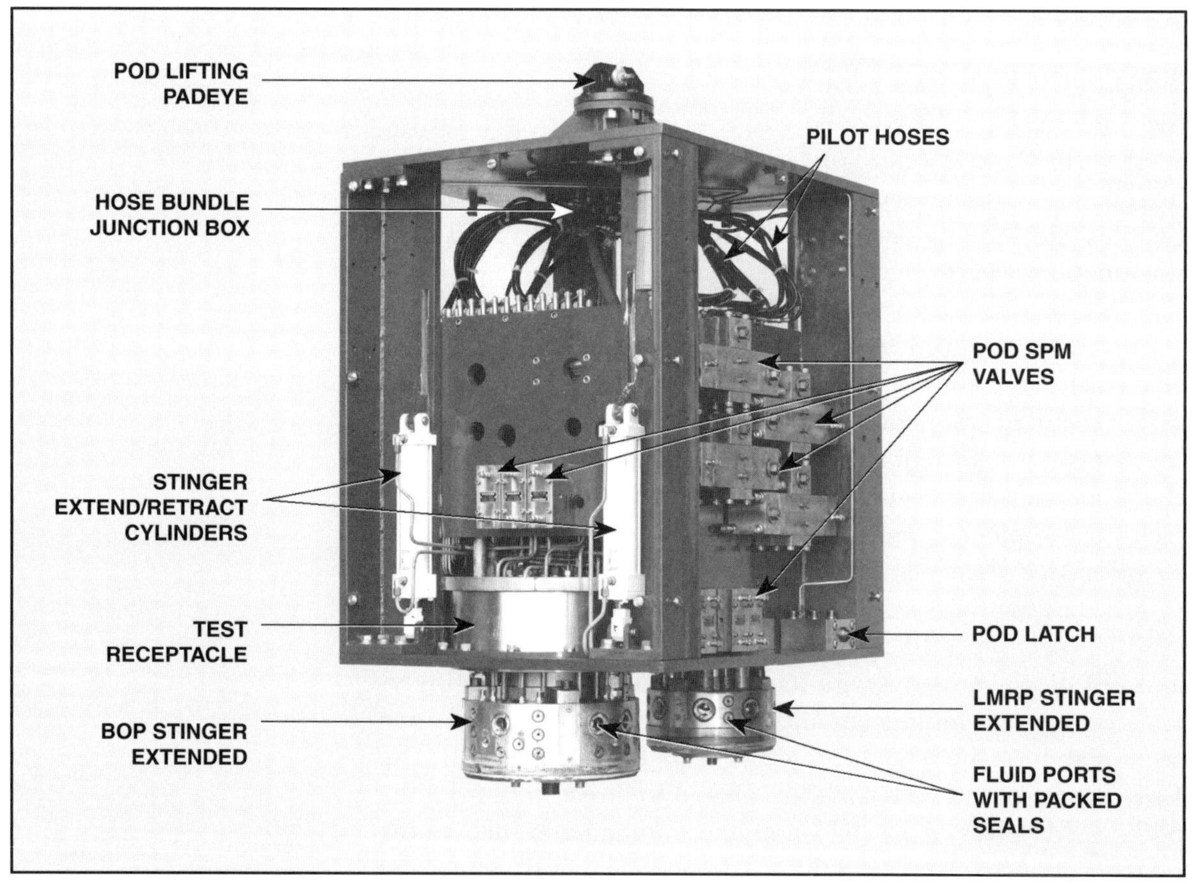

Figure 68. BOP and LMRP stingers extended by hydraulic cylinders (Courtesy of Cameron)

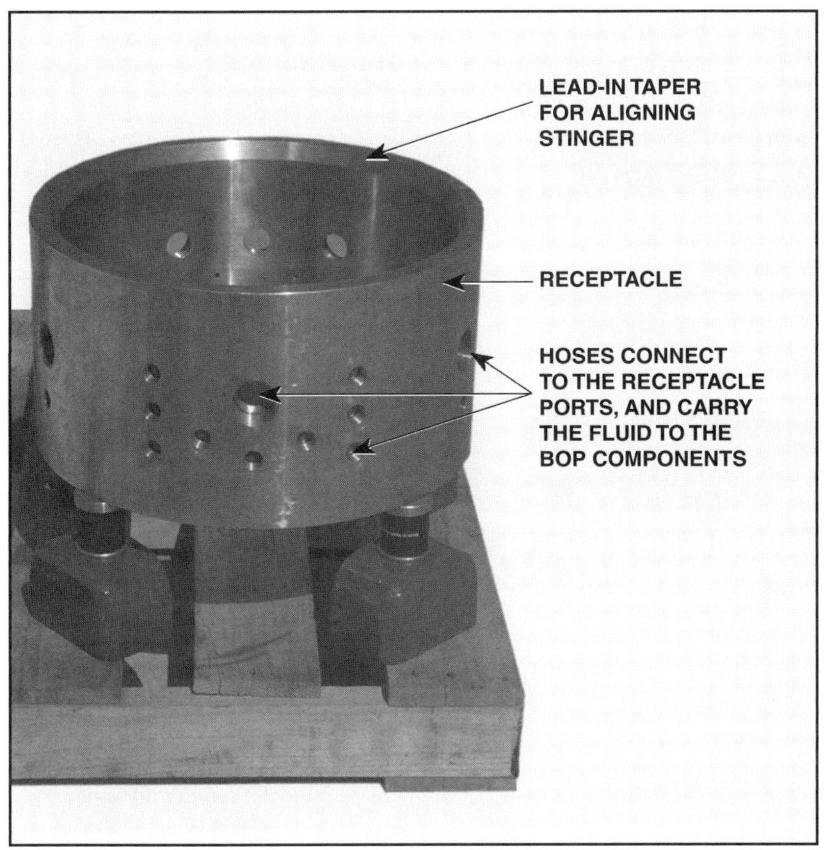

Figure 69. Receptacle for receiving BOP and LRMP stingers (Courtesy of Cameron)

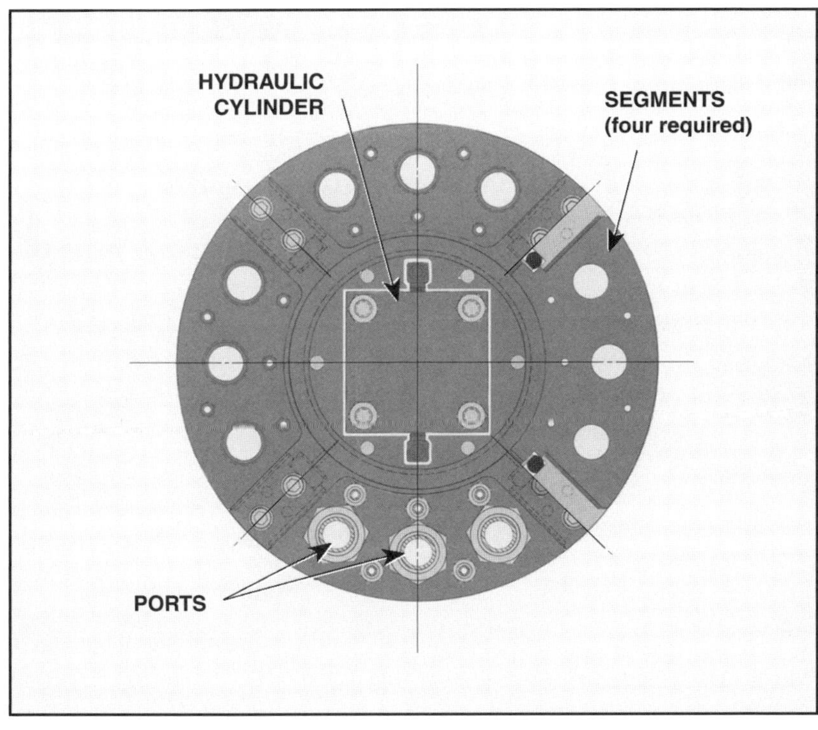

Figure 70. Top view of a stack pod stinger showing ports and segments (Courtesy of Cameron)

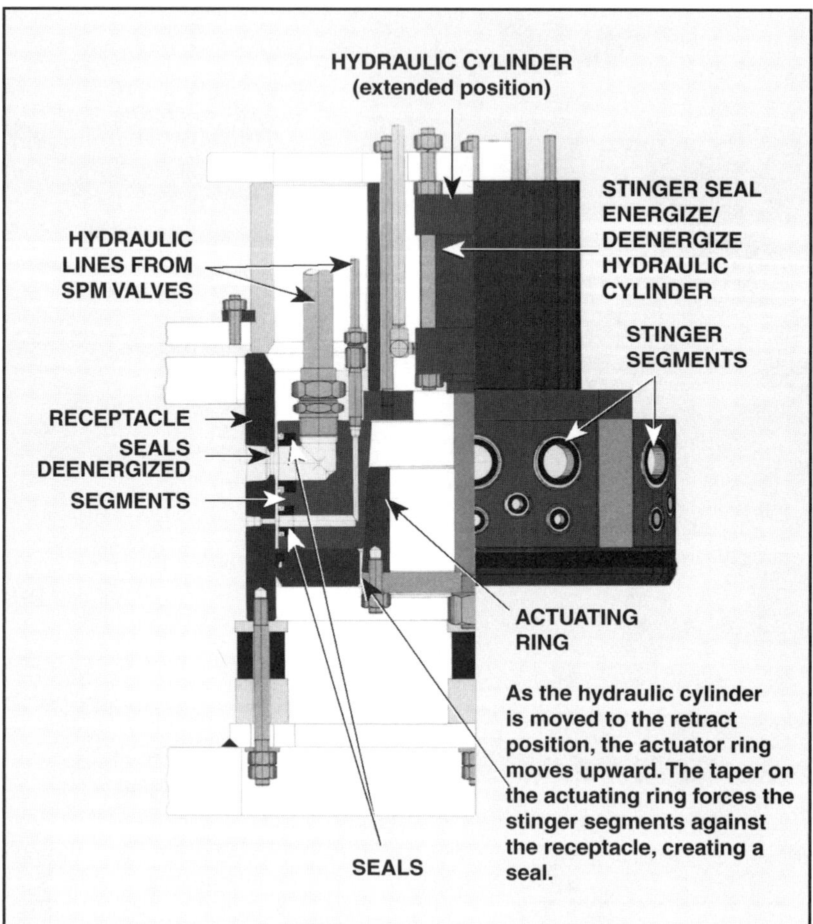

Figure 71. A stack pod stinger with segments in retracted position (Courtesy of Cameron)

prior to retracting the stinger to prevent seal damage (fig. 71). To energize the seals, a hydraulic cylinder is activated to force the stinger segments outward against the receptacle to create a seal. To deenergize the seals, the hydraulic cylinder is activated in the opposite direction, which removes the outward force from the stinger segments.

Pod Regulators

Accumulator supply pressure is routed to pod regulators that regulate the accumulator pressure from 3,000-psi (21,000-kPa) supply pressure to the BOP's normal working pressure. Usually, two pod regulators are provided; one is the BOP manifold regulator and the other is the annular BOP regulator. The regulated outlet port of the BOP manifold regulator supplies regulated pressure to the inlet ports of the BOP's SPM valves, including the ram preventers, wellhead and LMRP connectors, and subsea choke and kill valves. The regulated outlet port of the annular regulator supplies regulated pressure to the inlet ports of the annular preventer SPM valves.

The BOP manifold regulator is a hydraulically operated pilot valve. It has a pressure ratio of 1 to 1—that is, the regulated output pressure is equal to the hydraulic pilot pressure applied. So, if a pilot pressure of 1,500 psi (10,500 kPa) is applied to the regulator, the output pressure to the SPM valves will be 1,500 psi (10,500 kPa).

The pilot signal to the BOP manifold pod regulator is sent from the pilot regulator on the surface accumulator unit through a pilot hose in the hose bundle. At the surface accumulator unit, a tee is connected to the pilot line, which, in turn, is connected to the BOP manifold's regulator gauge and pressure transducer. This arrangement provides a pressure readout on the BOP manifold's pressure gauge at the accumulator unit and at the remote panels.

Because the pilot signal sent from surface is variable from 0 to 3,000 psi (21,000 kPa), the output pressure from the pod regulators is also variable from 0 to 3,000 psi (21,000 kPa). A tee is connected to the regulated output manifold of the regulator, which, in turn, is connected to a pilot line. The output pressure travels back to the surface accumulator unit and is connected to the BOP manifold's read-back pressure gauge and pressure transducer. The read-back pressure is therefore an indication of the actual pressure that is being applied to the BOP components.

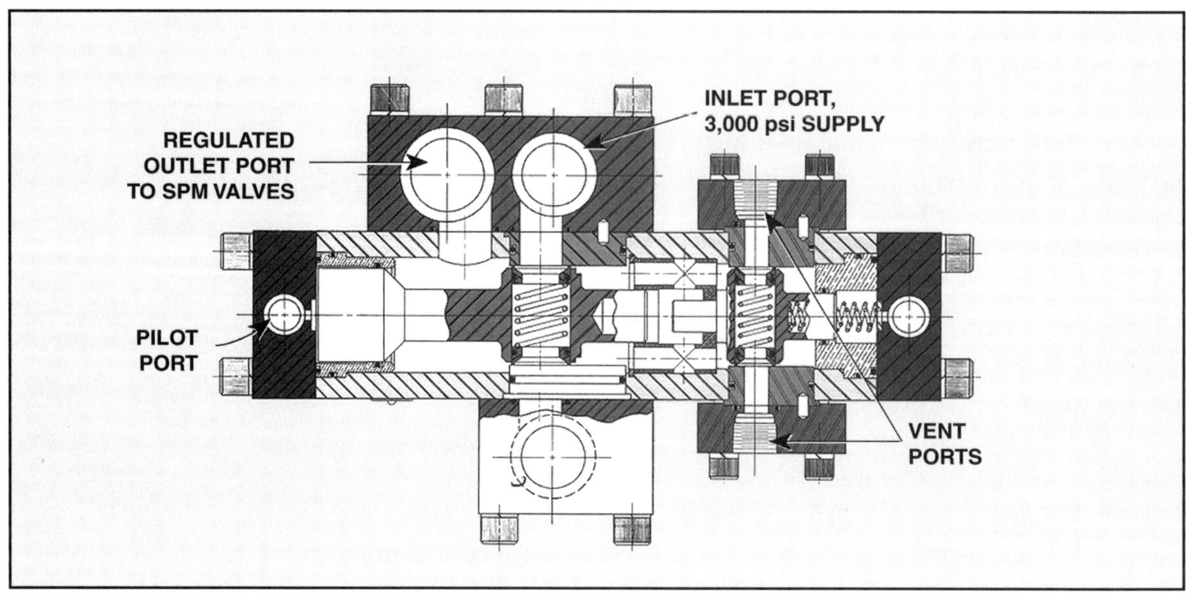

Figure 72. Pod annular regulator (Courtesy of Cameron)

The pod's annular regulator (fig. 72) operates the same as the BOP manifold's regulator. It has its own surface pilot regulator, accumulator gauges and transducers, and pilot lines.

SPM Valves

The SPM valves that are supplied by the pod regulators are hydraulically operated pilot valves. They are three-position, four-way valves that direct regulated pressure to the BOP components. The three positions are—

1. open, which allows regulated pressure to flow to the open side of the BOP component and vents the close side of the BOP component;
2. close, which allows regulated pressure to flow to the close side of the BOP component and vents the open side of the BOP component; and
3. block, which blocks the flow of regulated pressure and simultaneously vents the open and close sides of the BOP component.

A three-position, four-way SPM valve is installed for each three-position BOP component. The three positions of a BOP component are—

1. pressured open and close chamber vented,
2. pressured close and open chamber vented, and
3. open and close chambers both vented.

Pilot signals from the ¼-in. (6.35-mm) pilot-control valves mounted on the surface accumulator unit operate the SPM valve for each three-position component. One ¼-in. (6.35-mm) pilot-control valve is available for each SPM valve.

As an example of SPM operation, assume that the operator wishes to close a ram preventer. In this case, the operator places the surface ¼-in. (6.35-mm) pilot-control valve in the close position. With this action—

- pilot fluid at 2,000 psi (14,000 kPa) flows through the valve to the pilot line and through the hose bundle to the close pilot port on the pod's SPM valve.
- the open pilot signal simultaneously vents from the SPM valve's open pilot port through the hose bundle and the ¼-in. (6.35-mm) pilot-control valve's vent port.
- the ¼-in. (6.35-mm) pilot-control valve goes to the close position.
- the SPM valve receives the close pilot signal and shifts position, allowing the regulated pressure from the BOP's manifold regulator to flow to the close side of the ram operator. The fluid in the open side of the ram operator vents through the SPM valve's vent port.

To open the ram, the operator places the surface ¼-in. (6.35-mm) pilot-control valve in the open position. With this action—

- pilot fluid at 2,000 psi (14,000 kPa) flows through the valve to the pilot line and through the hose bundle to the open pilot port on the pod's SPM valve. The close pilot signal simultaneously vents from the SPM valve's close pilot port, through the hose bundle, and through the ¼-in. (6.35-mm) pilot-control valve's vent port.
- the SPM valve receives the open pilot signal and shifts position, allowing the regulated pressure from the BOP manifold regulator to flow to the open side of the ram operator. The fluid in the close side of the ram operator vents through the SPM valve's vent port.

To vent the ram, the operator places the surface ¼-in. (6.35-mm) pilot-control valve in the block position. With this action—

- pilot fluid is blocked at the pilot valve and both the open and close pilot signals vent through the ¼-in. (6.35-mm) pilot-control valve's vent port.
- the SPM valve shifts to the center position, which blocks the regulated pressure. With regulated pressure blocked, both the open and close sides of the ram vent through the vent port of the SPM valve.

SPM valves on rams, annular BOPs, and connectors on the BOP operate in the same way. However, the size of the SPM varies, depending on the fluid volume required to operate a particular component. For example, large volume components such as annular BOPs, require a 1½-in. (38.1-mm) SPM valve. Rams normally use a 1-in. (25.4-mm) SPM, while connectors require a ¾-in. (19.05-mm) SPM. Choke and kill valves normally use ½- or ¼-in. (12.7- or 6.35-mm) SPM valve. Figure 73 shows a typical ¼-in. (6.35-mm) SPM valve.

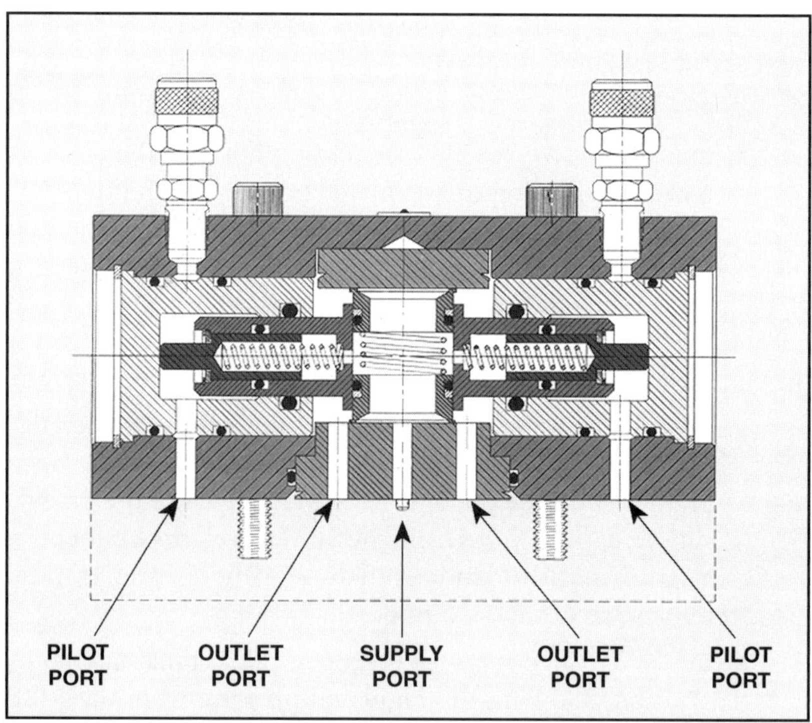

Figure 73. Typical ¼-inch (6.35-mm) SPM valve (Courtesy of Cameron)

Pod Latch Mechanism

On retrievable pods, a latching mechanism is provided to allow remote connection and disconnection of the pod from the LMRP base plate. The pod is retrieved and lowered by means of a retrieving line that is attached to a pod lifting padeye on the top plate of the pod frame. Two hydraulic cylinders are located on the pod frame's lower plate. When activated, the pistons are forced under brackets located on the LMRP base plate and lock the pod in place. The pistons are spring loaded; so, by simply venting hydraulic pressure, the springs retract the pistons clear of the brackets and the pod can be lifted clear of the LMRP.

Shuttle Valves and Hoses

Flexible hoses and shuttle valves between the upper and lower female pod receptacles supply fluid from the SPM valve outlets to the BOP inlet ports. The flexible hoses must have the same pressure rating as the maximum working pressure of the BOP components. The shuttle valves have two inlet ports and one outlet port. The valves are constructed to allow the connection of two fluid supplies to a single inlet port. The shuttle valve's outlet port is connected directly to the BOP component's inlet port, and the hoses from the yellow and blue pods are connected to the shuttle valve's inlet ports.

An internal spool or piston separates the two inlets. Thus, when supply is from the yellow pod, the internal spool seals off the blue supply port. On the other hand, when supply is from the blue pod, the internal spool slides across the valve and seals off the yellow supply port. The shuttle valve provides the redundancy between the two pods.

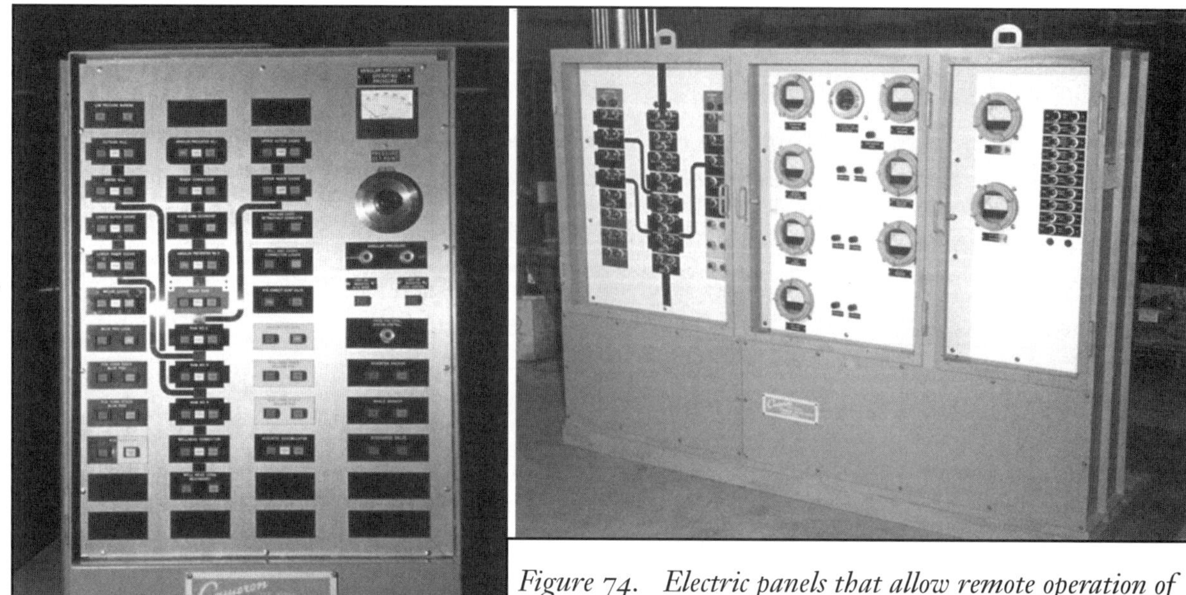

Figure 74. Electric panels that allow remote operation of BOP components (Courtesy of Cameron)

Remote Operation

Usually, two remote electric panels are provided to enable remote operation of all the BOP components (fig. 74). One panel is mounted at the driller's station, and the other at a safe location on the rig, usually in the toolpusher's office. The electrical power supply to the remote panels is either 24 or 120 volts, and a battery backup system is installed in case of power failure.

The remote panels have a graphic display of the BOP stack. Lights indicate the status of the BOP components, and each BOP component has an operating pushbutton. The panels also have a fluid flowmeter and pressure gauges. Normally, green lights indicate components in their normal drilling position—for example, rams open, annular BOPs open, and connectors locked. Red lights indicate a component that is not in normal drilling position.

Solenoid Valves

The solenoid valves are housed in an explosion-proof junction box at the accumulator unit. Rig air supplies them. When a BOP component is operated remotely by pressing a button on either of the remote panels, this action opens the solenoid, which allows rig air pressure to flow through the valve to the pneumatic cylinder connected to the ¼-in. (6.35-mm) pilot valve handle. The pneumatic cylinder then shifts the valve to the desired position. The solenoids can make the ¼-in. (6.35-mm) pilot valves open, close, or vent.

Pressure Switches

The pressure switches are housed in an explosion-proof junction box at the accumulator unit and are energized by hydraulic pressure. When the ¼-in. (6.35-mm) pilot valve sends a pilot signal to an SPM valve, a tee connected to the pilot line supplies the pilot signal to the pressure switch. Pilot pressure energizes the switch, which completes an electrical circuit, and illuminates the corresponding indicator light on both remote panels.

Pressure Transducers

The pressure transducers are housed in an explosion-proof junction box at the accumulator unit and convert hydraulic pressure into voltage. The voltage is sent from the transducer to the meters on the remote panels to give readouts of all system pressures.

Accumulator Sizing

API standards and government regulations address the issue of the required fluid volumes that should be stored in accumulators. Most systems are built to API recommendations, but they are often altered to meet local government regulations. API's publication entitled *Blowout Prevention Equipment Systems for Drilling Operations, RP 53* states:

> As a minimum requirement, closing units of subsea installations should be equipped with accumulator bottles with sufficient volumetric capacity to provide the usable fluid volume (with pumps inoperative) to open and close the ram preventers and one annular preventer and retain a 50% reserve.

Usable fluid volume is the volume of fluid recoverable between the accumulator operating pressure and 200 psi (1,400 kPa) above the precharge pressure. For example, assume that a BOP stack has four ram preventers and an annular preventer. Each ram requires 19.40 gal (73.44 L) to close it and 16.70 gal (63.22) to open it. The annular requires 61.37 gal (232.31 L) to close it and 47.76 gal (180.79 L) to open it. Table 3 shows the volume calculation in tabular form. A stack with four ram preventers and one annular preventer requires that the accumulator hold 380.30 gal (1,439.60 L) of operating fluid.

Assume that the rig is using 10-gal (37.5-L) accumulator bottles. Remember: each bottle contains not only operating fluid, but also nitrogen. So, both nitrogen volume and operating fluid volume must be accounted for. To make the calculation, Boyle's Law must be used. Boyle's Law states that at a constant temperature the volume of a gas varies inversely with its pressure. In other words, the higher the pressure on a gas, the less is its volume.

Table 3
Example Fluid Operating Volumes

Component	Closing Volume, gal (L)	Opening Volume, gal (L)	Number in System	Total gal (L)
Ram BOPs	19.40 (73.44)	16.70 (63.22)	4	144.40 (546.61)
Annular BOP	61.37 (232.31)	47.76 (180.79)	1	109.13 (413.10)
			Total for all BOPs	253.53 (959.71)
			+ 50%	126.77 (479.86)
				380.30 (1,439.60)

To calculate the number of 10-gal accumulator bottles required to hold 380.3 gal of operating fluid, use this equation—

$$V_3 = \frac{V_r}{\frac{P_3}{P_2} - \frac{P_3}{P_1}} \qquad \text{Eq. 4}$$

where

V_3 = total volume of nitrogen and fluid, gal (L)
V_r = total usable fluid required, gal (L)
P_1 = maximum working pressure, psi (kPa)
P_2 = minimum working pressure, psi (kPa)
P_3 = nitrogen precharge pressure, psi (kPa).

As an example, use the 380.3-gal total determined in Table 3 and assume that P_1 is 3,000 psi, P_2 is 1,200 psi, and P_3 is 1,000 psi. Therefore—

$$V_3 = \frac{380.3}{\frac{1{,}000}{1{,}200} - \frac{1{,}000}{3{,}000}}$$

$$= \frac{380.3}{0.833 - 0.333}$$

$$= \frac{380.3}{0.5}$$

$$V_3 = 760.6 \text{ gal.}$$

Because this example accumulator requires 760.6 gal of operating fluid and uses 10-gal bottles, simply divide 760.6 by 10 to determine that 77 ten-gal bottles are needed (760.6 ÷ 10 = 76.6 = 77). (Note that in the SI system, equation 4 is also used; however, equivalent volumes in L and equivalent pressure in kPa should be employed.)

Subsea Accumulators

Accumulators can be mounted on the subsea BOP and LMRP for many reasons. For example, subsea accumulators can decrease BOP closing times because operating fluid does not have to travel as far. Also, they can provide dedicated fluid storage for a high-pressure shear ram function. Whatever the purpose of the subsea accumulators, they must have sufficient capacity for the intended function. Where subsea accumulators provide improved response times, they are normally sized to have sufficient capacity to close one annular preventer. Where used for high-pressure shearing, they are sized to close only the shear rams at full working pressure. Equation 4 can be used for calculating subsea accumulator capacity; however, the effects of hydrostatic pressure must be included in the calculation by increasing the 1,000-psi (7,000 kPa) surface precharge pressure by the equivalent hydrostatic pressure.

Backup Systems

Backup systems ensure that certain BOP components can be operated even if the control system completely fails. In some cases, the backup is a remotely operated vehicle (ROV). The ROV ties into and operates certain functions such as closing the shear rams or unlocking the LMRP connector. The ROV has a ported hydraulic stab, which mates with a receptacle in the hydraulic circuit. The ROV locates the hydraulic stab in the receptacle and pumps hydraulic fluid into the circuit to operate the selected component.

Acoustic backup systems are more complex. Typically, they consist of a miniature control pod with a sufficient number of SPM valves to operate the selected BOP components and an acoustic receiver. Acoustic signals are sent to a subsea receiver, which sends the commands to operate the components.

Multiplex Electronic Control Systems

Deep water limits the efficiency of straight hydraulic control systems. The deeper the water, the longer it takes the pilot signal to travel through the pilot hose. And, the longer it takes the pilot signal to travel, the longer it takes to operate the BOP component. Generally, operators set maximum operating times—for example, 45 seconds for ram preventers and 60 seconds for annular preventers. The timing of the closing sequence starts when the actuating button is pushed on the remote panel and ends when the BOPs are fully closed; so, the timing includes the pilot signal time. Moreover, as the pilot signal builds pressure, the pilot hose expands volumetrically and this expansion increases the signal time.

To overcome delays, multiplex electronic control (MUX) systems are used in water depths greater than 5,000 ft (1,500 metres). Multiplex systems reduce pilot signal time and therefore the overall closing time of the preventers. Two subsea electronic MUX packages, labeled yellow and blue, are mounted directly to the hydraulic control pods (fig. 75).

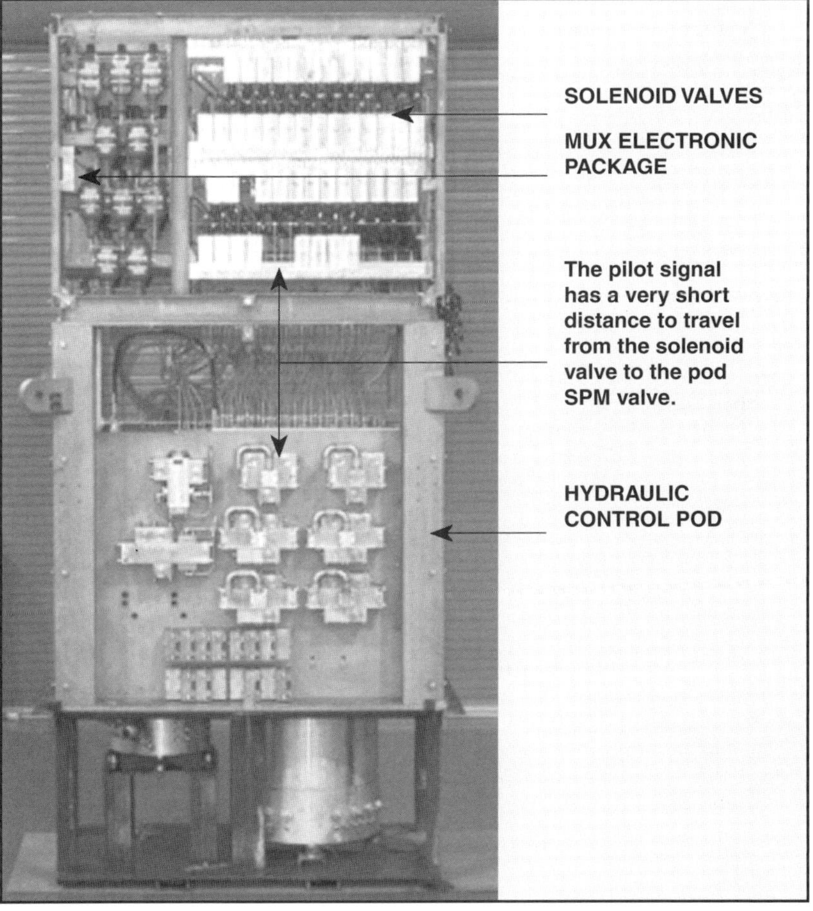

Figure 75. Electronic MUX package mounted on a hydraulic control pod. (Courtesy of Cameron)

Surface electronics transmit electronic command signals through the cable reels to the subsea MUX electronics packages, which decode and deliver the commands to solenoid valves mounted in the subsea MUX package. The solenoids are supplied with 3,000-psi (21,000-kPa) pilot pressure from the hydraulic system. When energized, the solenoids deliver the 3,000-psi (21,000-kPa) pilot pressure to the hydraulic control pod's SPM valves and regulators. Consequently, the hydraulic pilot signal travels a very short distance to the SPM valve or regulator from the solenoid (see fig. 74). This short travel distance, combined with the fact that the electronic signal's travel time is only a few milliseconds, makes the multiplex system much quicker than a straight hydraulic system.

The hydraulic control pod on a MUX system is identical to the straight hydraulic control pod. A MUX pod uses the same SPM valves and subsea regulators as the straight hydraulic system. The primary difference is that the MUX system provides the pilot signal very quickly.

The accumulator unit on a MUX system is merely a pumping and accumulator storage unit; no pilot signals travel from the surface to the pods. Consequently, no pilot manifold is required at the accumulator unit; moreover, no hose reels are needed. Hydraulic conduit lines mounted on the riser deliver accumulator pressure to the subsea components. Normally, two conduit lines are installed: one for the yellow pod supply and the other for the blue pod supply. At the surface, the operator selects them by means of conduit select valves. The conduit lines are usually manufactured from stainless steel to prevent corrosion, have an ID of at least 2 in. (50.8 mm), and are rated to 5,000 psi (35,000 kPa). The 2-in. ID and 5,000-psi accumulator supply pressure provides extremely fast fluid supply to the subsea control pods.

Conduit Manifold

The conduit lines terminate at a conduit manifold on the LMRP. The conduit manifold is made up of several valves that direct the two conduit supplies to the control pods. When one pod is selected, the other is vented.

A conduit flush valve allows the conduit lines to be flushed out prior to selecting a pod to prevent any contaminants from entering the pods. A pod flowmeter in the supply line to each pod measures fluid consumption in each pod. A crossover in the conduit manifold allows the yellow conduit line to supply the blue pod, and vice versa, should one line fail.

Another valve in the conduit manifold supplies the 5,000-psi (35,000-kPa) accumulator supply pressure to the yellow and blue solenoid packages through a check valve and a regulator. The regulator reduces the 5,000 psi (35,000 kPa) to 3,000 psi (21,000 kPa) prior to the pressure reaching the solenoids. Accumulator bottles on the solenoid packages store pilot pressure at 3,000 psi (21,000 kPa). The check valve in the supply line coming from the conduit manifold prevents pressure drops in the accumulator circuit, which can affect solenoid pressure. Once 3,000-psi (21,000-kPa) pilot pressure supplies the solenoids, the solenoids simply have to be energized to send the pilot signals to the SPM valves and regulators in the hydraulic pod.

Electronics

Two independent electrical power supplies (EPSs) are provided in a MUX system (fig. 76). The two electrical systems can power the entire electronic control system individually, and can provide seamless changeover should either supply fail. Usually, two surface control panels—a driller's panel and a toolpusher's panel—are provided. Either can fully and independently control the subsea control system.

Further, two identical subsea electronics modules (SEMs) are mounted in the subsea electronic MUX control pods. Each SEM has redundant dual electronic systems and each system communicates via a modem with communication controllers in the surface electronic system.

Redundant data pathways transmit commands from the panels to the controllers. Each control unit receives changes in status to ensure they are updated continuously and have identical data. The interface between the surface panel controllers and the SEMs is provided by modem cards that convert all commands into a secure message format for transmission to the SEMs in the subsea control pods.

Programmable logic controllers (PLCs) in the surface communication system are located at the accumulator unit, the driller's panel, toolpusher's panel, and distribution cabinets. The controllers consist of a central processing unit (CPU), communication controllers, power supply modules, and interface boards. All components are redundant so that, should a failure occur in one system, another system can take over. Also, the busses between the controllers are redundant.

Commands initiated at either the driller's or toolpusher's panel are sent through the busses to the communication controller and are received by the other controllers. All controllers update their information and indicate that a command has been executed at another panel. The subsea link communication controller sends the command by modem to the control pod electronics through the cable reels.

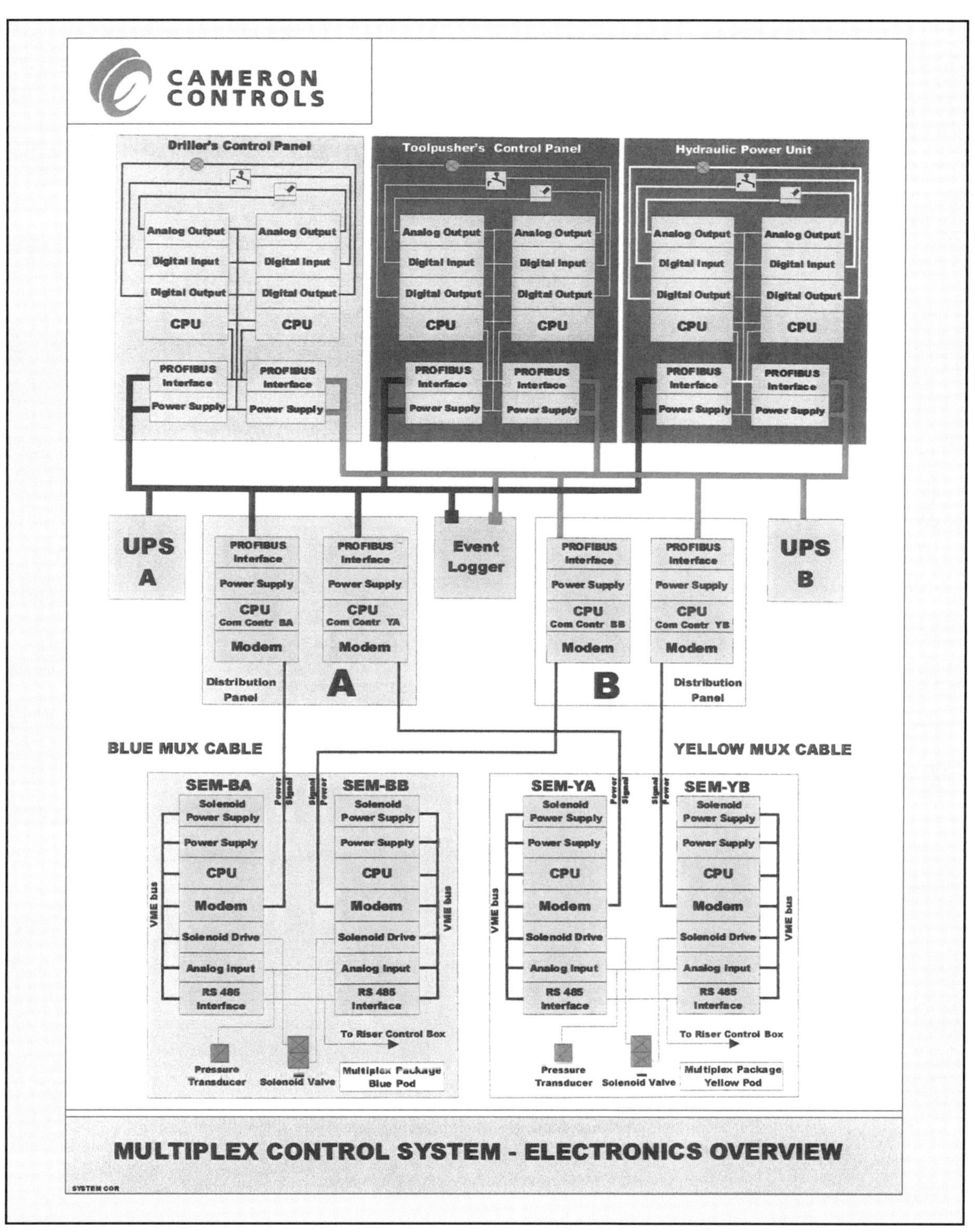

Figure 76. Multiplex control system (Courtesy of Cameron)

Cable Reels

The cable reels (fig. 77) can store over 10,000 ft (3,000 m) of electric cable. The cable links the power and electronic equipment on the surface with the subsea power and electronic equipment. Slip rings on the cable reel's shafts allow power and electronic data to continue to be transmitted to the subsea components even when crew members run or retrieve the BOP assembly. A pneumatic motor drives the reels, and a cable level-wind mechanism properly spools the cables onto the drum. Crew members should handle the cables with care because they are easily kinked and damaged.

Figure 77. Cable reels on a MUX system (Courtesy of Cameron)

Subsea Electronic Modules (SEMs)

The SEMs are housed in nitrogen-filled vessels in each of the control pods. They consist of a PLC; a modem, which communicates with the surface link communication controllers; and a 230-volt alternating current (VAC) power supply, which transformer-rectifiers convert to 5 volts direct current (VDC) for the PLC and 24 VDC for the solenoids.

Figure 78 shows an SEM removed for testing. The electronic components convert the serial signals received from the surface to discreet commands for each solenoid valve. When the solenoid receives the command from the SEM, it energizes and opens to allow pilot fluid at 3,000 psi (21,000 kPa) to flow to its corresponding

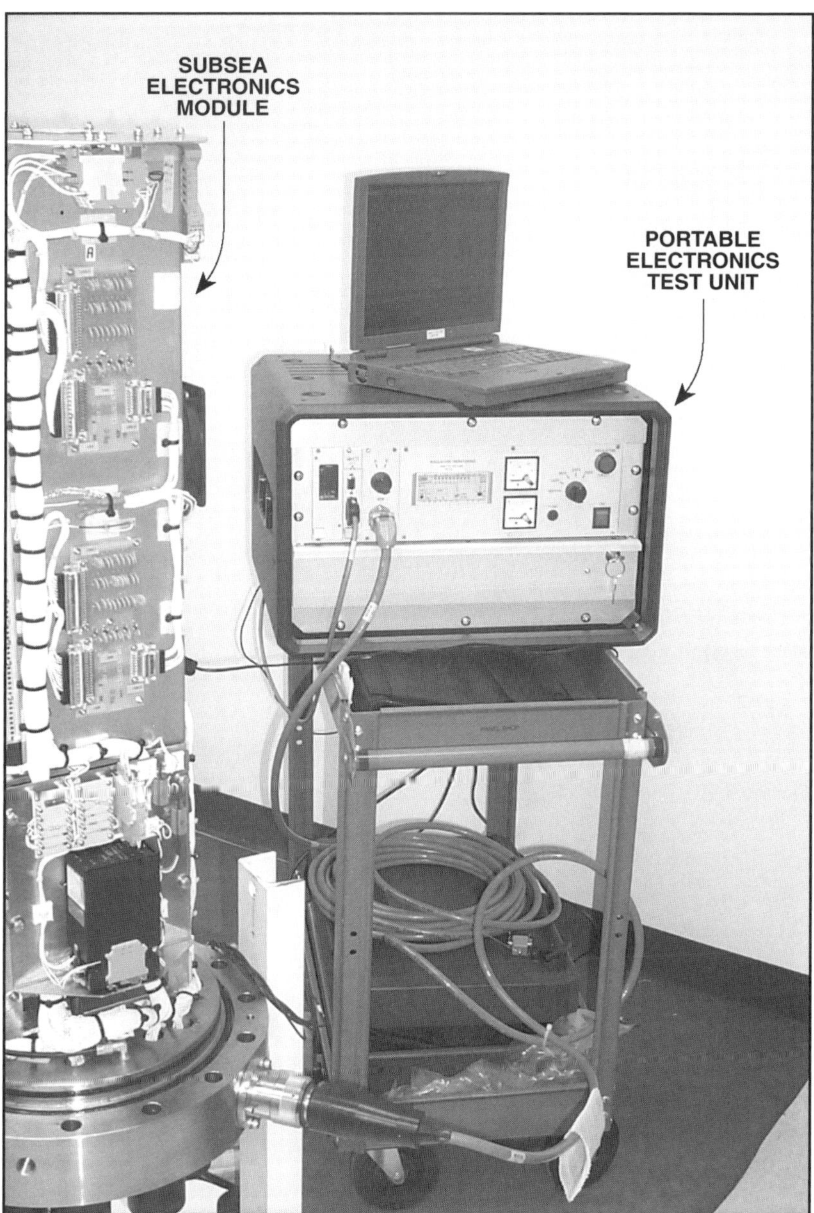

Figure 78. SEM removed from housing for testing (Courtesy of Cameron)

SPM valve or regulator pilot port. Figure 79 is a schematic of the basic circuit used to close an annular BOP.

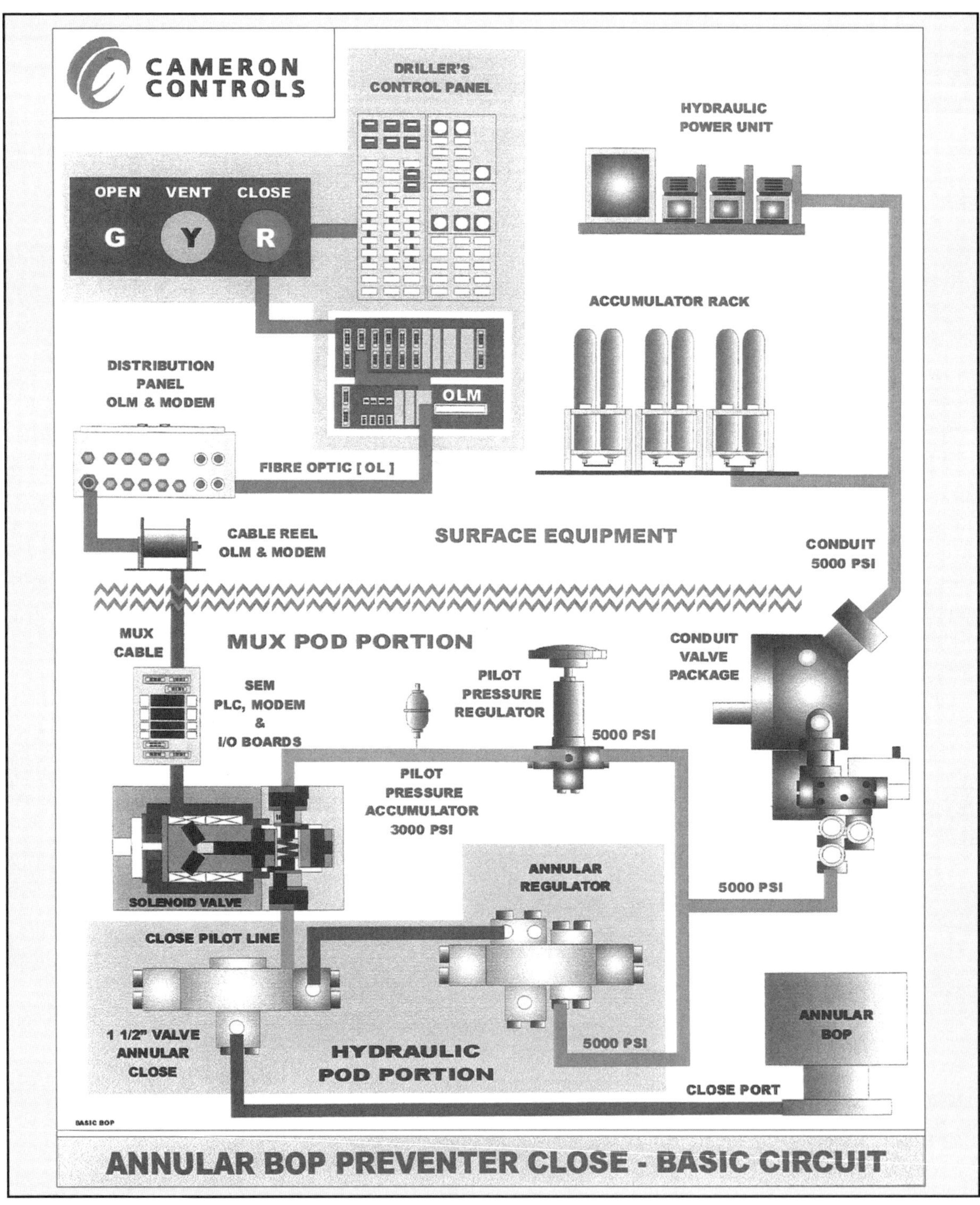

Figure 79. An open SEM under test (Courtesy of Cameron)

To summarize—

Hydraulic control systems—

- operate the BOPs.
- vent hydraulic fluid from the opposite side of the operated component.
- are indirect pilot-operated systems.
- are either straight hydraulic type or the multiplexed type.

Straight hydraulic control systems are rated to 3,000-psi (21,000-kPa) working pressure and deliver the hydraulic pilot signals from the surface through pilot hoses to the valves and regulators mounted in the subsea control pods.

The multiplexed system is usually rated to 5,000-psi (35,000-kPa) working pressure and sends electronic signals from the surface to subsea-mounted solenoid valves that then send the pilot signals to the valves and regulators.

The straight hydraulic system consists of—

- the surface accumulator unit, or HPU.
- the hose bundles, hose reels, and hose reel control panels.
- the subsea control pods.
- remote electric panels.

Multiplex (MUX) systems—

- are used in water depths greater than 5,000 ft (1,500 metres) to overcome operating delay time.
- use surface electronic systems to transmit electronic command signals through the cable reels to the subsea MUX electronics packages, which decode and deliver the commands to solenoid valves mounted in the subsea MUX package.

Glossary

A

abnormal pressure *n*: strictly speaking, pressure in a formation that is less than or more than the pressure to be expected at a given depth. However, in the field, abnormal pressure is often considered to be pressure only that is higher than that which is expected at a given depth. Normal pressure increases approximately 0.465 pounds per square inch per foot of depth or 10.5 kilopascals per metre of depth (this value may be slightly more or slightly less than 0.465 psi or 10.5 kilopascals, depending on a particular area). In general, however, normal formation pressure at 1,000 feet is 465 pounds per square inch; at 1,000 metres it is 10,500 kilopascals. See *pressure gradient*.

accumulator *n*: on a drilling rig, an assembly of devices such as bottles, control valves, pumps, and hydraulic fluid reservoirs that stores hydraulic fluid under pressure and provides a way for personnel to operate (open and close) the blowout preventers. See *blowout preventer, blowout preventer control unit.*

accumulator bottle *n*: a bottle-shaped steel cylinder located in a blowout preventer control unit to store nitrogen and hydraulic fluid under pressure (usually at 3,000 pounds per square inch). The fluid operates the blowout preventers. See *blowout preventer control unit.*

acoustic backup system *n*: in offshore drilling from a floating drilling rig using a subsea blowout preventer stack, devices that send acoustic signals to a subsea receiver to operate the blowout preventer (BOP) components. The system consists of a miniature control pod with several subsea pilot manipulated (SPM) valves to operate the selected BOP components. The system is used when the hydraulically operated system fails.

actuator *n*: a device that activates or puts into motion a process or an action by use of pneumatic, hydraulic, or electronic signals; for example, a valve actuator opens or closes a valve.

American Petroleum Institute (API) *n*: an oil trade organization (founded in 1920) that is the leading standardizing organization for oilfield drilling and producing equipment. Headquartered in Washington D.C., it publishes materials concerning exploration and production, petroleum measurement, marine transportation, marketing, pipelining, refining, safety and fire protection, storage tanks, valves, training, health and environment, policy, and economic studies. Address: 1220 L Street NW; Washington, DC 20005; (202) 682-8000; www.api.org.

annular blowout preventer *n*: a large valve, usually installed above the ram blowout preventers, that, when closed, forms a seal in the annular space between the pipe and the wellbore or, if no pipe is present, in the wellbore itself. Compare *ram blowout preventer*.

annular piston *n*: in drilling from floating offshore drilling rigs using a subsea blowout preventer stack and a Cameron model HC hydraulic connector, which is a device that connects the lower marine-riser package (LMRP) to the subsea blowout preventer stack, a solid cylinder (a piston) that completely surrounds the locking segments in the connector and which provides an additional locking force. The additional locking force available through the annular piston increases the pressure rating to 15,000 psi (105,000 kPa). See *hydraulic connector, lower marine-riser package*.

annular space *n*: 1. the space that surrounds a cylindrical object within a cylinder. 2. the space around a pipe in a wellbore, the outer wall of which may be the wall of either the borehole or the casing; often termed the annulus.

annulus *n*: see *annular space*.

antiextrusion plate *n*: in a ram blowout preventer, a flat steel piece (a plate) on the ram block that prevents rubber from being forced out of the area between the drill pipe and the ram block when wellbore pressure is applied, and which feeds reserve rubber into contact with the drill pipe as rubber is lost because of wear. See *ram, ram blowout preventer*.

API *abbr*: American Petroleum Institute.

assembly *n*: a group of components that make up a mechanism, machine, or similar device.

automatic flooding valve *n*: in a subsea marine riser, a device that prevents collapse of the riser from the surrounding water pressure when the riser is inadvertently emptied of drilling mud. The valve is incorporated into a riser joint and, when it senses an increase in pressure from the seawater surrounding the empty riser, it opens to allow seawater into the riser to prevent its collapse.

axial load *n*: the vertical weight an object carries that extends perpendicularly into the object.

axial tension *n*: vertical upward force (tension) placed on an object.

B

backup system *n*: a set of devices in an assembly or in a process that can take over the operation of the assembly or process should the devices normally used to operate the assembly or process fail.

ball joint *n*: in drilling from floating offshore drilling rigs that employ a subsea blowout preventer system, a device mounted between the annular preventer and the riser adapter on the lower marine-riser package (LMRP) to prevent excessive bending forces from being exerted on the marine riser, the LMRP, and the blowout preventer components. It is a forged-steel ball and socket containing a cylindrical neck extension with a riser adapter attached at the top of the neck. Compare *flex joint*.

barite *n*: barium sulfate, $BaSO_4$; a mineral frequently used to increase the weight or density of drilling mud. Its relative density is 4.2 (i.e., it is 4.2 times denser than water). See *barium sulfate, mud*.

barite sack *n*: a container (a sack) of powdery material (barite) that weighs about 100 pounds (about 45 kilograms). Barite is a dense mineral with a specific gravity of about 4.2—it is over four times heavier than water. On drilling rigs, barite (barium sulfate, or $BaSO_4$) can be supplied in heavy-duty paper sacks. In drilling from floating drilling rigs, sacks of barite are sometimes placed on top of the temporary guide base before it is run to the seafloor. The sacks provide additional weight to the guide base and ensure that it makes firm contact with the seafloor. Also, barite is often used to increase the density, or weight, of drilling mud. See *barite*, *temporary guide base*.

barium sulfate *n*: a chemical compound of barium, sulfur, and oxygen ($BaSO_4$), often used to increase the density of drilling mud. Also called barite.

base plate *n*: 1. in drilling from floating rigs, a flat steel piece (a plate) that is part of the riser support spider (the device that supports riser pipe in the rotary table as it is being run or retrieved). The base plate supports the spider and fits into the rig's rotary table. 2. in drilling from floating rigs, a flat steel piece (a plate) at the top of the lower marine-riser package (LMRP) on top of which the control pods are attached and which supports the pods on the LMRP. See *lower marine-riser package*, *riser support spider*.

bending load pressure *n*: the force created on an object by applying the weight of another object so that the weight bends or tends to bend the first object.

bending moment *n*: a force that lateral movement (bending) creates on an object.

bending stress *n*: in offshore drilling from floating rigs, a force placed on a marine-riser system and subsea blowout preventer stack that originates from movements of the water and from the weight of the upper components. Wind, waves, and currents act on the equipment to cause it to bend as does the weight of components in the system.

blind end *n*: the end of a device, such as the end of a cylinder used in a riser tensioning system, which is capped and completely closed, or blinded.

blind ram *n*: an integral part of a blowout preventer, which serves as the closing element on an open hole. Its ends do not fit around the drill pipe but seal against each other and shut off the space below completely. See *ram*.

blind ram preventer *n*: a blowout preventer in which blind rams are the closing elements. See *blind ram*.

blind-shear ram preventer *n*: see *shear ram*.

blowout *n*: an uncontrolled flow of gas, oil, or other well fluids into the atmosphere or into an underground formation. A blowout may occur when formation pressure exceeds the pressure applied to it by the column of drilling fluid and rig crew members fail to take steps to contain the pressure. Before a well blows out, it kicks; thus a kick precedes a blowout. See *kick*.

blowout preventer (BOP) *n*: one of several valves installed at the wellhead to stop (prevent) the escape of pressure either in the annular space between the casing and the drill pipe or in open hole (i.e., hole with no drill pipe) during drilling or completion operations. Blowout preventers on land rigs are normally located beneath the rig at, or slightly below, the land's surface; on jackup or platform rigs, at the water's surface; and on floating offshore rigs, on the seafloor. See *annular blowout preventer*, *ram blowout preventer*.

blowout preventer (BOP) control unit *n*: a device that stores hydraulic fluid under pressure in special containers and provides a method to open and close (operate) the blowout preventers quickly and reliably. Usually, compressed air and hydraulic pressure provide the opening and closing force in the unit. See *blowout preventer*. Also called an accumulator.

blowout preventer (BOP) mandrel *n*: in drilling from offshore floating rigs, a metal fitting connected to the top of the annular preventer in the subsea blowout preventer stack. The mandrel has a mating-and-seal profile to match that of the lower marine-riser package's (LMRP's) connector. The connector normally has the same pressure rating as the LMRP's annular preventer.

blowout preventer (BOP) manifold regulator *n*: in drilling from floating drilling rigs using a subsea blowout preventer, a special device that controls (regulates) the pressure to the inlet ports of a subsea blowout preventer's subsea pilot manipulated (SPM) valves, which, in turn, operate the ram preventers, wellhead and lower marine-riser package (LMRP) connectors, and subsea choke and kill valves.

blowout preventer (BOP) stack *n*: an assembly of blowout preventers placed on top of each other (stacked one on top of the other), which typically consists of one or two annular preventers and three, four, or more ram preventers. See *annular preventer*, *ram preventer*.

bolt *n*: a metal fastener made from a threaded pin or rod, usually with a square or hexagonal head at one end, which is inserted through holes in assembled parts and secured by a mated nut that is tightened by applying torque.

bolted-flange riser connector *n*: in drilling from floating drilling rigs using a subsea blowout preventer and riser system, a riser connector in which bolts are screwed into threaded holes in a flange welded on the end of the riser. The bolts are tightened to strict specifications with special torque wrenches to avoid failure of the connection. See *riser pipe*.

bonnet *n*: 1. the part of a valve that packs off and encloses the valve stem. 2. on a ram blowout preventer, the housing that surrounds the rams and contains the operating piston and rod.

BOP *abbr*: blowout preventer.

BOP assembly *n*: see *blowout preventer stack*.

BOP control system *n*: see *blowout preventer control unit*.

BOP stack *n*: the assembly of blowout preventers installed on a well. See *blowout preventer stack*.

break out *v*: 1. to unscrew one section of pipe from another section, especially drill pipe while it is being withdrawn from the wellbore. During this operation, the tongs are used to start the unscrewing operation. 2. to separate, as gas from a liquid or water from an emulsion.

buoyancy *n*: the apparent loss of weight of an object immersed in a fluid. If the object is floating, the immersed portion displaces a volume of fluid the weight of which is equal to the weight of the object.

buoyant riser joint *n*: in drilling from offshore floating rigs in deep water, a marine-riser joint to which has been added syntactic foam modules. The syntactic foam modules, which contain many hardened spheres of air that are embedded in a rugged foam-like material, are attached to the riser joint. The spheres provide buoyancy. Buoyant riser joints are required when drilling in deep water (depths beyond 4,000 feet or 1,200 metres) to relieve stress on the riser tensioning system. Compare *slick riser joint*. See *riser pipe*, *syntactic foam*.

C

cable reel *n*: in offshore drilling from floating drilling rigs, a large spool around which is wrapped the electric cable that provides the link between the surface power and electronics and the subsea power and electronics on the lower marine-riser package and the subsea blowout preventer stack. Because most subsea blowout preventers have two operational systems in case one fails, two cable reels are usually provided. A typical cable reel can store over 10,000 feet (3,000 metres) of cable.

cam-actuated running tool (CART) *n*: in drilling from floating rigs that employ a subsea blowout preventer system, a tool used to run the permanent guide base. It allows the drill pipe to be disconnected from the permanent guide base after it lands in the temporary guide base. When the drill pipe is rotated, the cams move into a position that allows the pipe to be removed from the base. See *permanent guide base*, *temporary guide base*.

cam ring *n*: a part of a hydraulic connector that attaches the blowout preventer (BOP) stack to the wellhead and the lower marine-riser package (LMRP) to the BOP stack. The cam ring is connected to hydraulic pistons in the connector and the ring's inside diameter has a 4-degree taper. When hydraulic locking pressure is applied, the pistons pull the cam ring downward over the locking dogs or segments and the 4-degree taper forces the dogs or segments inward to lock on the wellhead's profile.

CART *abbr*: cam-actuated running tool.

casing *n*: steel pipe placed in an oil or gas well to prevent the wall of the hole from caving in, to prevent movement of fluids from one formation to another, and to improve the efficiency of extracting petroleum if the well is productive. Most casing joints are manufactured to specifications established by API, although non-API specification casing is available for special situations. Casing manufactured to API specifications is available in three length ranges. A joint of range 1 casing is 16 to 25 feet (4.8 to 7.6 metres) long; a joint of range 2 casing is 25 to 34 feet (7.6 to 10.3 metres) long; and a joint of range 3 casing is 34 to 48 feet (10.3 to 14.6 metres) long. The outside diameter of a joint of API casing ranges from 4½ to 20 inches (114.3 to 508.0 millimetres). Casing is made of many types of steel alloy, which vary in strength, corrosion resistance, and so on.

casing hanger *n*: a circular device with a frictional gripping arrangement of slips and packing rings used to suspend casing from a casinghead in a well.

casinghead *n*: a heavy, flanged steel fitting connected to the first string of casing. It provides a housing for slips and packing assemblies, allows suspension of intermediate and production strings of casing, and supplies the means for the annulus to be sealed off. Also called a spool.

casing-shear rams *n*: usually high-capacity shear rams capable of shearing drill collars and casing strings that are installed below the blind-shear rams in a BOP stack. Casing-shear rams can be used in addition to the blind-shear rams to shear pipe. See *shear ram*.

casing string *n*: the entire length of all the joints of casing run in a well. Most wells have more than one string of casing, which are run one inside the other as drilling progresses. Generally, wells have a length, or string, of surface casing, intermediate casing, and production casing, although a great deal of variation occurs according to the well's depth, the formations it penetrates, and other factors. See *casing, intermediate casing string, production casing, surface pipe*.

cavity *n*: a hollow area or a hole within the body of a piece of equipment, a device, or a rock.

central processing unit (CPU) *n*: in computer technology, the electronic device that contains the circuits necessary to receive, interpret, and carry out instructions received from peripherals such as keyboards, monitors, hard drives, and the like.

check valve *n*: a valve that permits fluid to flow in one direction only. If the gas or liquid starts to reverse, the valve automatically closes, preventing reverse movement. Sometimes referred to as a one-way valve.

choke *n*: a special valve with an opening that restricts the flow of fluids. Chokes are used to control the rate of flow of the drilling mud out of the hole when the well is closed in with the blowout preventer and a kick is being circulated out of the hole.

choke and kill system *n*: an assembly of lines (pipes), valves, and operating devices (a system) on a subsea blowout preventer that provides a conduit for drilling mud to be pumped into the well through a kill line and a conduit for fluids exiting from the well to flow through a choke line to the surface. See *choke line, kill line*.

choke and kill valve *n*: a device that, when opened, allows drilling mud to be pumped into the well and drilling and other fluids returning from the well to be directed to the choke manifold on the surface. See *choke line, kill line*.

choke line *n*: a pipe, or conduit, usually installed on outlets below the ram blowout preventers, through which fluids in the annulus can flow to the surface and through the choke manifold when valves to it are open. On floating offshore rigs, flow may be directed either down the annulus or up the annulus; on most land rigs, flow is only upward.

choke manifold *n*: an arrangement of piping and special valves, called chokes. In drilling, mud is circulated through a choke manifold when the blowout preventers are closed.

closing pressure *n*: in a blowout preventer stack, the amount of pressure supplied by the blowout preventer control unit that is required to close a preventer. See *operating pressure*.

closing ratio *n*: the mathematical relation (the ratio) between the pressure in the hole and the operating-piston pressure needed to close the rams of a blowout preventer. The closing ratio represents the difference in area between the part of the ram preventer's operating piston that is affected by wellbore pressure, and the part of the piston that closing pressure acts on.

Coflexip hose *n*: a brand name of a flexible, high-strength, steel-braided, high-pressure hose that conducts fluids on a drilling rig and other petroleum installations. Many manufacturers make this type of hose. It is very strong and has high crush, abrasion, and fatigue resistance. It is used for choke and kill lines, cementing lines, surface blowout preventer control lines, injection lines, and other applications.

completion *n*: to finish work on a well and bring it to productive status. See *well completion.*

compression load *n*: the weight (load) placed on a string of pipe, such as riser or casing, that results when force from above presses down on the pipe.

conductor casing *n*: generally, but not always, the first string of casing in a well. It may be lowered into a hole drilled into the formations near the surface and cemented in place; it may be driven into the ground by a special pile driver (in such cases, it is sometimes called drive pipe); or it may be jetted into place in offshore locations. Its purpose is to prevent the soft formations near the surface from caving in and to conduct drilling mud from the bottom of the hole to the surface when drilling starts. Also called conductor pipe, drive pipe.

conductor housing *n*: in offshore drilling from floating rigs, a special fitting welded to the top of the conductor casing that is profiled to fit into the permanent guide base. It connects the conductor casing to the guide base. See *conductor casing, permanent guide base.*

conduit *n*: a pipe, line, or channel through which fluids flow.

conical funnel *n*: in drilling from floating offshore rigs, one of four cone-shaped devices (funnels) mounted on top of the temporary guide base that guides the blowout preventer stack and lower marine-riser package into place on the base. See *blowout preventer stack, lower marine-riser package, temporary guide base.*

connection *n*: 1. a section of pipe or fitting used to join pipe to pipe or to a vessel. 2. a place in electrical circuits where wires join. 3. *v*: the action of adding a joint of pipe to the drill stem as drilling progresses.

connector *n*: see *hydraulic connector.*

control panel *n*: in various types of systems, the switches and devices to start, stop, measure, monitor, or signal what is taking place within the system.

control pod *n*: see *hydraulic control pod.*

counterbalance *n*: a force or influence equally counteracting another. For example, drilling fluids of the correct height and weight, or density, should exert enough pressure to counterbalance formation pressure *v*: to put in balance; to balance, equalize, stabilize, or steady.

CPU *abbr*: central processing unit.

D

deepwater drilling *n*: a relative term that refers to offshore drilling operations in deep oceans or seas—currently, about 5,000 feet (1,500 metres) and deeper. It presents a number of special problems related to water depth.

deepwater riser system *n*: a marine riser used in deepwater drilling operations and that may be equipped with special devices such as automatic flooding valves, which open to admit seawater into the riser and prevent collapse of the riser when accidental evacuation of mud in a riser occurs. Also, some of the riser joints may be equipped with buoyant riser modules. See *buoyant riser joint*, *riser pipe*.

design load *n*: the most stressful combination of weight or other forces a mechanical system or device is built to sustain.

differential *n*: the variation in quantity or degree between two measurements or units. For example, the pressure differential across a choke is the difference between the pressure on one side and that on the other.

diverter *n*: in offshore drilling, an assembly of devices used to direct fluids flowing from a well away from the drilling rig. When a kick is encountered at shallow depths, the well often cannot be shut in safely because shutting it in on a shallow formation may create pressures high enough to fracture (break down) the formation. Therefore, a diverter is used. When activated, it allows well fluids to flow through a side outlet to a line (pipe) that carries the well fluids a safe distance away from the rig. A diverter contains a packing element, flow-line seals, and lock-down mechanisms and is run and retrieved with a special handling tool made up on riser pipe.

diverter assembly *n*: see *diverter*.

diverter system *n*: see *diverter*.

dog *n*: any of various simple devices for holding, gripping, or fastening, such as a hook, rod, or a spike with a ring, claw, or lug at the end.

dog-type riser connector *n*: a riser connector that has actuator screw assemblies in the box end. Tightening the screws with pneumatic torque wrenches causes tapered locking dogs to engage mating tapers in the pin end. The application of the correct torque to the actuator screw, combined with the taper geometry of the pin and locking dogs, preloads the connection to the design load rating. Spring-loaded locks prevent vibration from backing out the actuator screws. The actuator screws do not support any of the riser connection's load. See *hydraulic connector*.

driller's console *n*: a metal cabinet on the rig floor containing the controls that the driller manipulates to operate various components of the drilling rig.

driller's panel *n*: see *driller's console*.

driller's position *n*: the area immediately surrounding the driller's console.

drill floor *n*: also called rig floor or derrick floor. See *rig floor*.

drilling fluid *n*: circulating fluid, one function of which is to lift cuttings out of the wellbore and to the surface. It also serves to cool the bit and to counteract downhole formation pressure. Although a mixture of barite, clay, water, and other chemical additives is the most common drilling fluid, wells can also be drilled by using air, gas, water, or oil-base mud as the drilling mud. Also called circulating fluid, drilling mud. See *mud*.

drilling rig *n*: see *rig*.

drilling template *n*: see *temporary guide base*.

drill pipe *n*: seamless steel or aluminum pipe made up in the drill stem between the kelly or top drive on the surface and the drill collars on the bottom. During drilling, it is usually rotated while drilling fluid is circulated through it. Drill pipe joints are available in three ranges of length: 18 to 22 feet (5.49 to 6.71 metres), 27 to 30 feet (8.23 to 9.14 metres), 38 to 45 feet (11.58 to 13.72 metres). The most popular length is 27 to 30 feet.

drill pipe float *n*: a check valve installed in the drill stem that allows mud to be pumped down the drill stem but prevents flow back up the drill stem. Also called a float.

drill pipe safety valve *n*: see *drill stem safety valve*.

drill stem *n*: all members in the assembly used for rotary drilling from the swivel to the bit, including the kelly, the drill pipe and tool joints, the drill collars, the stabilizers, and various specialty items. Compare *drill string*.

drill stem safety valve *n*: a special valve normally installed below the kelly. Usually, the valve is open so that drilling fluid can flow out of the kelly and down the drill stem. It can, however, be manually closed with a special wrench when necessary. In one case, the valve is closed and broken out, still attached to the kelly, to prevent drilling mud in the kelly from draining onto the rig floor. In another case, when kick pressure inside the drill stem exists, the drill stem safety valve is closed to prevent the pressure from escaping up the drill stem. Also called lower kelly cock, mud saver valve.

drill string *n*: the column, or string, of drill pipe with attached tool joints that transmits fluid and rotational power from the kelly to the drill collars and the bit. Often, especially in the oil patch, the term is loosely applied to include both drill pipe and drill collars. Compare *drill stem*.

DTL *abbr*: dynamic tension limit.

dual sheave *n*: in a riser tensioner system, a pair of pulleys (sheaves) that is mounted on the blind end of the system's cylinder and over which the tension lines are reeved.

dynamic loading *n*: the exerting of a force, such as weight, on an object with continuous movement; cyclic stressing.

dynamic tension limit (DTL) *n*: in a riser tensioning system, the maximum allowable pressure, multiplied by the effective hydraulic area, divided by the number of line parts. It is determined from the equation $D_{tl} = P_a \times A_{cyl} \div N_{lp}$ where D_{tl} = dynamic tension limit, P_a = maximum allowable system operating pressure, A_{cyl} = effective hydraulic area, and N_{lp} = number of line parts.

E

elastomer *n*: a synthetic rubber made from polymers that has the elastic properties of natural rubber; packers (sealing elements) in blowout preventers and downhole packers are often made of elastomer. The term is formed by combining part of the word elastic and part of the word polymer—i.e., elast(ic) and poly(mer).

elbow *n*: a fitting that allows two pipes to be joined at an angle of less than 180 degrees, usually 90 or 45 degrees.

electrical junction box *n*: 1. a housing (box) into which electrical wires from two or more sources enter, join, and exit to other destinations. 2. a housing (box) mounted on an accumulator unit to provide an interface for remote electrical panels and an accumulator unit's pod select valve, pilot valves, and various system pressures; normally two electrical junction boxes are provided in case one fails.

electrically-driven triplex pump *n*: in a blowout preventer's accumulator system, a reciprocating pump with three pistons that electric motors operate.

engineering analysis *n*: in offshore drilling, the examination and study of such factors as tensile loads, bending stresses, maximum operational water depth, buoyancy requirements, surface tension, and vessel response for the purpose of designing the best subsea marine riser and blowout prevention system.

F

fail-safe *adj*: capable of compensating automatically and safely for a failure, as of a mechanism or power source.

fatigue characteristic *n*: in metallurgy, a measure of the trait of a metal, such as steel, to withstand repeated, or cyclic, stress without failure.

fixed-bore ram block *n*: in a ram blowout preventer, a steel block with elastomer surfaces that seal around drill pipe of a specific size. Besides forming a seal around drill pipe to confine well pressure in the annulus, they also support the load of the drill string—that is, the drill string can be hung off on them. Compare *variable bore ram*. See *hang off, pipe ram*.

flange *n*: a projecting rim or edge (as on pipe fittings and openings in pumps and vessels), usually drilled with holes to allow bolting to other flanged fittings.

flapper valve *n*: a type of check valve in a pipe or a line that has a hinged closure mechanism (a flapper) and that allows fluid flow in one direction but shuts it off in the other direction.

flexible hose *n*: a type of tube or pipe that is bendable (flexible) so that repeated movements do not cause it to break; frequently used in subsea blowout preventer systems to conduct operating fluid from the accumulator on the surface to operating devices on the marine-riser pipe.

flex joint *n*: on floating offshore drilling rigs using a subsea blowout preventer, a device mounted between the annular preventer and the riser adapter on the lower marine-riser package (LMRP). Flex joints bend laterally to prevent excessive bending forces from being exerted on the marine riser and the LMRP and BOP components. (A bending moment is a force lateral movement creates on an object.) A flex joint typically allows 10 degrees of offset from vertical. The riser adapter can be connected to the top of the flex joint's neck by a flange, a hub, or by welding. Compare *ball joint*.

floating offshore drilling rig *n*: a type of mobile offshore drilling unit (MODU) that floats and is not in contact with the seafloor (except possibly with anchors) when it is in the drilling mode. Floating units include barge rigs, drill ships, and semisubmersibles. See *mobile offshore drilling unit*.

float valve *n*: see *drill stem safety valve*.

flowmeter *n*: an instrument that monitors, measures, or records the amount of fluid moving through a pipe or other container.

flow-rate sensor *n*: a device mounted in the mud return flow line that detects the speed (flow rate) of the mud flow and sends a signal to an instrument on the rig floor and other rig locations to warn of a change in the return flow rate. An increase in flow may indicate that the well has kicked, while a decrease may indicate a loss of returns.

fluid pressure *n*: a force exerted on a pipe, borehole, or vessel in which a fluid (a gas or a liquid) is confined. The fluid may be at rest or moving. Fluid pressure may be measured in force per unit area, such as pounds per square inch (psi) or as a derived unit from force per unit area, such as pascals (Pa), which is derived from newtons per square meter (N/m^2).

fluid reservoir *n*: 1. a geological formation that contains hydrocarbon fluids (oil and gas). See *reservoir*. 2. a container (a reservoir) on a blowout preventer control unit (an accumulator) that holds hydraulic fluid used to operate the blowout preventers. On offshore floating rigs, where subsea blowout preventers are often employed, two reservoirs are usually mounted on the accumulator unit's skid. One reservoir is used to store the fluid concentrate and one is used to store the mixed fluid. See *blowout preventer control unit*.

formation fluid *n*: fluid (such as gas, oil, or water) that exists in a subsurface rock formation.

formation pressure *n*: the force exerted by fluids in a formation, recorded in the hole at the level of the formation with the well shut in. Also called reservoir pressure or shut-in bottomhole pressure.

freestanding *adj*: of or pertaining to objects in the wellbore that do not contact the sides of the hole or the riser pipe, but which may be suspended from equipment on the surface.

full-bore *adj*: of a valve, fitting, or other object with an opening placed in a pipe, or line, which, when fully open, matches (or nearly matches) the diameter of the pipe in which the valve, fitting, or object is mounted.

G

galvanic cell *n*: a natural battery (a cell) that occurs when two dissimilar metals are put into contact with each other. The difference in electrical potential between the two metals causes current to flow from one metal to the other, and, in the process, take away minute amounts of one of the metals and deposit it on the other.

galvanic corrosion *n*: a type of corrosion that occurs when a small electric current flows from one piece of metal equipment to another. It is particularly prevalent when two dissimilar metals are present in an environment in which electricity can flow (as two dissimilar joints of tubing in an oil or gas well).

galvanic protection *n*: corrosion prevention employed at the point where dissimilar metals come into contact and at which point corrosion could occur because of the small current created when dissimilar metals touch. The protection interrupts the current flow and thus prevents corrosion. See *galvanic cell*, *galvanic corrosion*.

gas-cut mud *n*: a drilling mud that contains entrained formation gas, giving the mud a characteristically fluffy texture. When entrained gas is not released before the fluid returns to the well, the weight or density of the fluid column is reduced. Because a large amount of gas in mud lowers its density, gas-cut mud must be treated to reduce the chance of a kick.

gas flow *n*: in well control, the streaming (the flow) of gas from the well when a formation containing gas is penetrated and is not contained by the drilling mud or well-control equipment.

gasket *n*: any material (such as paper, cork, asbestos, or rubber) used to seal two essentially stationary surfaces.

gate valve *n*: an opening and closing device (a valve) that employs a slab of metal (a gate) with a hole in it that is moved up or down within the valve's body. When the hole in the gate is positioned opposite the opening in the valve, fluid flows through the valve. When the solid part of the gate is positioned opposite the valve's opening, flow stops. Gate valves are not used to regulate the flow of fluids through them—that is, they are either fully open or fully closed.

gimbal *n*: a mechanical frame that permits an object mounted in it to remain in a stationary or near-stationary position regardless of movement of the frame. Gimbals are often used offshore to counteract undesirable wave motion.

glycol *n*: a group of compounds used to dehydrate gaseous or liquid hydrocarbons or to inhibit the formation of hydrates. Glycol is also used in engine radiators as an antifreeze. Commonly used glycols are ethylene glycol, diethylene glycol, and triethylene glycol. See *hydrate*.

go-devil *n*: a device that is lowered into the borehole of a well for various purposes such as enclosing surveying instruments, detonating downhole explosive devices, and the like. *v*: to drop or pump a device down the borehole, usually through drill pipe or tubing.

gooseneck *n*: 1. the curved connection between the rotary hose and the swivel or top drive. 2. any curved length of pipe that serves as a connection between one conduit to another. See *swivel*, *top drive*.

guide base *n*: see *permanent guide base*, *temporary guide base*.

guide funnel *n*: see *conical funnel*.

guideline *n*: one of usually four lines attached to the temporary guide base and permanent guide base to help position equipment (such as blowout preventers) accurately on the seafloor when a well is drilled offshore from a floating vessel. Guidelines are normally used to drill in water depths of 5,000 feet (1,500 metres) or less.

guideline tensioner *n*: a system of cables (wire ropes), pulleys (sheaves), and pressurized piston-and-cylinder assemblies that maintain an upward tension on the guidelines to keep them taut and to compensate for rig heave (up-and-down movement). See *guideline*.

H

hang off *v*: to close a ram blowout preventer around the drill pipe when the annular preventer has previously been closed to offset the effect of heave on floating offshore rigs during well-control procedures.

HC connector *n*: a brand of hydraulic connector in subsea blowout preventer systems used to connect by remote control the blowout preventer stack to the wellhead and the lower marine-riser package to the blowout preventer stack. It uses a large annular piston that completely surrounds locking segments, instead of the nine individual pistons. The additional locking force available through the annular piston increases the pressure rating to 15,000 psi (105,000 kPa) and improves the resistance to bending moments. See *hydraulic connector*.

H_4 connector *n*: a brand of hydraulic connector that allows the blowout preventer stack to be connected by remote control to the wellhead and to the lower marine-riser package. The H_4 connector has both a primary and secondary piston connected to a cam ring to provide locking and unlocking force. See *hydraulic connector*.

high-capacity torque wrench *n*: a heavy-duty torque wrench used to apply the required torque to flange bolts in the makeup of riser pipe. Correct torque must be used to ensure the correct preload on a riser connection; otherwise, separation at the flanges could lead to failure. See *riser pipe*.

high-pressure wellhead *n*: in drilling from floating drilling rigs, a subsea device that is installed on top of the casing to provide a foundation for the blowout preventer stack, to support and house subsequent casing strings, and to provide a seal and a locking arrangement between the surface casing and blowout preventer stack.

hose bundle *n*: in drilling from floating offshore drilling rigs that employ a subsea blowout preventer stack, a collection (a bundle) of several hoses enclosed in a heavy-duty protective cover. A hose bundle has several electrical and hydraulic control lines that run to the control pods on the subsea blowout preventers. When the driller or other person actuates a blowout preventer control on the rig at the surface, the control signal goes through the hose bundle to the control pod on the blowout preventer stack.

hose bundle reel *n*: on offshore floating rigs, a large spool, or reel, onto which the hose bundle is wrapped. See *hose bundle*.

hose storage reel *n*: see *hose bundle reel*.

HPU *abbr*: hydraulic pressure unit.

hubbed connection *n*: a fitting for joining two pipes or devices; a metal disk (a hub) on each end of the pipes or devices to be connected are put together and bolted to form a pressure-tight seal.

hydrate *n*: a hydrocarbon and water compound that is formed under reduced temperature and pressure in places where gas that contains water vapor occurs. For example, hydrates can form in gathering, compression, and transmission facilities for gas. They can also form in deepwater drilling where low temperatures occur on or near the seafloor and where gas containing water vapor may be encountered by the borehole. Hydrates can accumulate in troublesome amounts and impede fluid flow. They resemble snow or ice and decompose at atmospheric pressure. *v*: to enlarge by taking water on or in.

hydrate seal *n*: in drilling from floating rigs that use a subsea blowout preventer stack and marine-riser system, a groove in the lower body of a hydraulic connector that forms a seal with the outside of the wellhead.

hydraulic actuator *n*: a cylinder or fluid motor that converts hydraulic power into useful mechanical work; mechanical motion produced may be linear, rotary, or oscillatory.

hydraulic bonnet operating system *n*: in a ram preventer, the devices in the ram housing (the bonnet) that open or close the rams when activated with hydraulic pressure from the blowout preventer operating unit (accumulator).

hydraulic connector *n*: in drilling from offshore floating rigs, a device that attaches (connects) the blowout preventer stack to the wellhead and to the lower marine-riser package. It is controlled from the surface. Hydraulic connectors consist of a lower body, an upper body, a cam ring, locking dogs (or segments), a seal groove, and a hydraulic system. See *lower marine-riser package*.

hydraulic control pod *n*: a device used on floating offshore drilling rigs to provide a way to actuate and control subsea blowout preventers from the rig on the surface. Hydraulic and electrical lines from the rig enter the pods, through which fluid and electrical signals are sent to the preventer. Usually two pods, one yellow and one blue, are used, each to safeguard and back up the other. Also called blue pod, yellow pod.

hydraulic junction box *n*: on a blowout preventer control unit (accumulator), a housing (box) with several inlet and outlet ports. The inlet ports receive the accumulator's operating fluid; the outlets send the fluid to the yellow and blue control pods, which, in turn, operate the blowout preventers.

hydraulic pressure unit *n*: see *blowout preventer control unit*.

hydraulic pump *n*: a device that creates pressure on a fluid, usually special hydraulic fluid, to move the fluid.

hydraulic riser handling tool *n*: see *riser handling tool*.

hydraulic supply line *n*: piping that carries hydraulic operating fluid from a storage container to a device that is operated by hydraulic fluid—for example, the piping from an accumulator to a blowout preventer.

hydraulic-tensioner ring *n*: a device attached to the telescopic joint on a floating rig's marine-riser system that allows the telescopic joint to be run and retrieved without removing tensioner lines. The ring is hydraulically latched to the outer barrel as the joint is run, and hydraulically latched to the diverter housing as the joint is being retrieved.

Hydril *n*: the registered trademark of a prominent manufacturer of oilfield equipment, including annular and ram blowout preventers and casing and tubing joints.

hydropneumatic cylinder *n*: a component of the riser tensioning system on a floating drilling rig that contains hydraulic fluid and high-pressure air and opposes heave placed on the system by ocean movements. See *hydropneumatic riser-tensioning system*.

hydropneumatic riser-tensioning system *n*: the components (the system) that provide upward force (tension) on a marine riser's tensioner lines. The system consists of hydropneumatic cylinders and sheave assemblies, accumulators and system air-pressure vessels, a control panel and piping manifold, high-pressure air compressors, and standby air-pressure vessels.

hydrostatic pressure *n*: the force exerted by a body of fluid at rest. It increases directly with the density and the depth of the fluid and is expressed in many different units, including pounds per square inch or kilopascals. The hydrostatic pressure of fresh water is 0.433 pounds per square inch per foot (9.792 kilopascals per metre) of depth. In drilling, the term refers to the pressure exerted by the drilling fluid in the wellbore.

I

IBOP *abbr*: inside blowout preventer.

inside blowout preventer (IBOP) *n*: any one of several types of valve installed in the drill stem or in a top drive to prevent high-pressure fluids from flowing up the drill stem and into the swivel, standpipe, or atmosphere. Also called an internal blowout preventer.

intermediate casing string *n*: the string of casing set in a well after the surface casing but before production casing is set to keep the hole from caving and to seal off troublesome formations. In deep wells, one or more intermediate strings may be required. Sometimes called protection casing.

internal blowout preventer *n*: also called inside blowout preventer. See *inside blowout preventer.*

International System of Units (SI) *n*: a system of units of measurement based on the metric system, adopted and described by the Eleventh General Conference on Weights and Measures. It provides an international standard of measurement to be followed when certain customary units, both U.S. and metric, are eventually phased out of international trade operations. The symbol SI (Le Système International d'Unités) designates the system, which involves seven base units: (1) metre for length, (2) kilogram for mass, (3) second for time, (4) Celsius for temperature, (5) ampere for electric current, (6) candela for luminous intensity, and (7) mole for amount of substance. From these units, others are derived without introducing numerical factors.

isolation test plug *n*: a special blocking device (a plug) that is used in pressure testing a blowout preventer. The plug is run and landed in a high-pressure wellhead housing, where it forms a pressure-tight seal. This seal protects (isolates) the casing below the wellhead from the high pressures used to test the blowout preventers, which are located above the wellhead and casing.

J

joint *n*: in drilling, a single length (from 16 feet to 45 feet, or 5 metres to 14.5 metres, depending on its range length) of drill pipe, drill collar, casing, or tubing that has threaded connections at both ends. Several joints screwed together constitute a stand of pipe.

junction box *n*: a housing (a box) that provides a place for electrical, hydraulic, or pneumatic connections to be made from one part of the system to another. For example, a junction box on an accumulator provides a point for connecting (interfacing) the remote electrical panels, as well as the accumulator unit's pod select valve, pilot valves, and various system pressures.

K

kelly *n*: on drilling rigs that do not use a top drive to rotate the bit, a heavy steel tubular device, four- or six-sided, suspended from the swivel through the rotary table and connected to the top joint of drill pipe to turn the drill stem as the rotary table turns. It has a bored passageway that permits fluid to be circulated into the drill stem and up the annulus, or vice versa. Kellys manufactured to API specifications are available in four- or six-sided versions, are either 40 or 54 feet (12 to 16 metres) long, and have diameters as small as 2.5 inches (6 centimetres) and as large as 6 inches (15 centimetres).

kelly cock *n*: originally, a term that referred only to the heavy-duty valve made up between the kelly and the swivel, which, when closed, kept back-pressure that was flowing up the drill stem from reaching the swivel and rotary hose. Today, on rigs that use a kelly and rotary table system to rotate the drill stem and bit, two kelly cocks are often employed: one between the kelly and the swivel—the upper kelly cock—and the other between the kelly and the first joint of drill pipe—the lower kelly cock. The lower kelly cock is also called a drill stem safety valve because, when closed and the kelly is removed from the drill stem, it keeps mud from falling out of the kelly. In any case, when a high-pressure backflow occurs inside the drill stem, and the kelly is made up in the drill stem, either valve may be closed to keep pressure off the swivel and rotary hose. See *lower kelly cock*, *upper kelly cock*.

kick *n*: an entry of water, gas, oil, or other formation fluid into the wellbore during drilling, workover, or other operations. It occurs because the pressure exerted by the column of drilling or other fluid in the wellbore is not great enough to overcome the pressure exerted by the fluids in a formation exposed to the wellbore. If prompt action is not taken to control the kick, or kill the well, a blowout may occur. See *blowout*.

kick-out sub *n*: on a lower marine-riser package (LMRP), which is part of the subsea marine-riser system, a fitting to which the kill- and choke-line hoses are attached. The sub has an angle of about 45 degrees, which extends (kicks out) the hoses so that they can fit around the LMRP components.

kill fluid *n*: mud or other fluid in a wellbore whose weight, or density, creates pressure great enough to equal or exceed the pressure exerted by formation fluids.

kill line *n*: a pipe, or conduit, usually attached to openings in the blowout preventer stack below the ram blowout preventers, that allows drilling mud or other fluid to be pumped into the well to control a well when it is not possible to pump down the drill string.

kill mud *n*: see *kill fluid*.

L

landing shoulder *n*: on conductor casing, a recessed, machined area on the conductor housing or similar fitting on which a wellhead is placed (landed).

liner *n*: a string of pipe used to case open hole below existing casing. A liner extends from the setting depth up into another string of casing, usually overlapping about 100 feet (30 metres) above the lower end of the intermediate or the oil string. Liners are nearly always suspended from the upper string by a hanger device.

LMRP *abbr*: lower marine-riser package.

load *n*: in mechanics, the weight or pressure placed on an object.

Load King connector *n*: a brand of bolted-flange connector used in marine-riser systems to join (connect) riser pipe. This connector is designed for drilling in water depths of 7,000 feet (2,000 metres) or deeper, and has a tensile load rating of 3.5 million pounds (1.6 million kilograms). See *riser pipe*.

locking device *n*: on the telescopic joint of a floating offshore rig, a mechanism that secures (locks) the joint's inner barrel to its outer barrel in a fully retracted position. Locking the joint fully retracted allows for easier handling when it is being shipped, when it is being picked up on the rig, and when the blowout preventer stack is being run or retrieved.

locking dog *n*: on a lower marine-riser package's connector; a device with a tapered locking profile on its inside diameter that mates with a tapered locking profile on the outside diameter of the wellhead.

lock nut *n*: a special type of threaded fastener (a nut) that, when screwed down firmly against another nut or a washer, fastens (locks) firmly in place.

lost circulation *n*: the quantities of whole mud lost to a formation, usually in cavernous, fissured, or coarsely permeable beds. Evidenced by the complete or partial failure of the mud to return to the surface as it is being circulated in the hole. Lost circulation can lead to a blowout and, in general, can reduce the efficiency of the drilling operation. Also called lost returns.

lost returns *n pl*: see *lost circulation*.

lower kelly cock *n*: on rigs that use a kelly and rotary table system to rotate the bit and drill stem, a heavy-duty valve installed between the kelly and the first joint of drill pipe. Usually, the valve is open so that drilling fluid can flow out of the kelly and down the drill stem. When the mud pump is stopped, the valve can be manually closed with a special wrench to prevent pressurized fluids in the drill string from flowing into the kelly. When closed, the valve also prevents mud in the kelly from spilling onto the rig floor when the kelly is broken out of the drill string. Also called a drill stem safety valve, mud saver valve.

lower marine-riser package (LMRP) *n*: the equipment that attaches the bottom part of the marine riser to the subsea blowout preventer (BOP) stack. It includes the BOP connector, a flexible joint to compensate for side-to-side movement, and the marine-riser connector. See *riser connector, riser pipe, subsea blowout preventer.*

lower marine-riser package (LMRP) connector *n*: a hydraulically actuated device that is used to remotely connect the blowout preventer stack to the wellhead, and the lower marine-riser package to the blowout preventer stack.

M

make up *v*: 1. to assemble and join parts to form a complete unit (e.g., to make up a string of drill pipe). 2. to screw together two threaded pieces. Compare *break out*. 3. to mix or prepare (e.g., to make up a tank of mud). 4. to compensate for (e.g., to make up for lost time).

manifold *n*: 1. an accessory system of piping to a main piping system (or another conduit) that serves to divide a flow into several parts, to combine several flows into one, or to reroute a flow to any one of several possible destinations. 2. a pipe fitting with several side outlets to connect it with other pipes. 3. a fitting on an internal-combustion engine made to receive exhaust gases from several cylinders.

manifold valve *n*: a valve placed in a series of piping (a manifold). For example, in a choke manifold on a drilling rig, an adjustable choke, which is placed in the manifold, is one of several types of manifold valve that may be present.

marine-riser pipe *n*: see *riser pipe*.

marine-riser system *n*: see *riser pipe*.

marine-riser tensioning system *n*: an assembly of devices installed on a floating offshore drilling rig to maintain a constant upward force (tension) on the riser pipe, despite vertical motions (heave) caused by ocean movements.

material grade *n*: in oilfield equipment and tools, a measure of the equipment or tool's strength—that is, its ability to be bent, stretched, compressed, or loaded without deformation or breaking.

mechanical riser handling tool *n*: a special device used to lower and pull riser joints on a floating offshore drilling rig using a subsea marine-riser system. The tool duplicates a riser-joint connection, and is made up on the riser joint in the same way as another riser joint. The upper section of the handling tool is a standard drill pipe tool joint that is supported by the elevators during raising and lowering operations. See *riser pipe*.

miniconnector *n*: on floating offshore rigs using a marine-riser system and a subsea blowout preventer stack, a device used to connect the choke and kill lines from the blowout preventer to the lower marine-riser package's connector. A miniconnector has a pressure rating of 15,000 psi (105,000 kPa) and an inside diameter of 3 inches (76.2 millimetres). See *lower marine-riser package*.

mixed fluid reservoir *n*: on an accumulator (blowout preventer operating unit), one of two containers (reservoirs). One reservoir stores the fluid concentrate and the other stores the mixed fluid. Using two reservoirs doubles the storage capacity of the accumulator.

mobile offshore drilling unit (MODU) *n*: a drilling rig that drills offshore exploration and development wells. It floats on the surface of the water when being moved from one well site to another, but it may or may not float once drilling begins. Two basic types of mobile offshore drilling units are used to drill most offshore wildcat wells: bottom-supported offshore drilling rigs and floating drilling rigs. Bottom-supported units include jackups and floating rigs include semisubmersibles and drill ships.

Model 70 connector *n*: a brand of a hydraulic connector used in subsea blowout preventer stacks that uses individual hydraulic cylinders located in the connector's lower body to operate an actuator ring. The connector has nine cylinders, six of which are primary, and three of which are secondary.

MODU *abbr*: mobile offshore drilling unit.

monoethylene glycol *n*: a chemical (an antifreeze) that is injected into lines and vessels to prevent hydrates from forming in the lines and vessels where hydrocarbon gases and water vapor are present at low temperatures. See *hydrate*.

MPL *abbr*: multiposition lock.

mud *n*: the liquid circulated through the wellbore during rotary drilling and workover operations. In addition to its function of bringing cuttings to the surface, drilling mud cools and lubricates the bit and the drill stem, protects against blowouts by holding back subsurface pressures, and deposits a mud cake on the wall of the borehole to prevent loss of fluids to the formation. Although it originally was a suspension of earth solids (especially clays) in water, the mud used in modern drilling operations is a more complex, three-phase mixture of liquids, reactive solids, and inert solids. The liquid phase may be fresh water, diesel oil, or crude oil and may contain one or more conditioners. See *drilling fluid*.

mud booster line *n*: in drilling from floating offshore drilling rigs that use a subsea blowout preventer stack, a line (pipe) sometimes provided on riser joints to increase the return velocity of the mud in the riser. It helps move heavy cuttings up the riser when normal pump circulation is inadequate. The mud booster line is connected to a pump on the rig that enables the driller to circulate drilling fluid from the surface to the bottom of the riser.

mud column *n*: the borehole when it is filled or partially filled with drilling mud.

mud-gas separator *n*: a device that removes gas from the mud coming out of a well when a kick is being circulated out.

mud pit *n*: originally, an open pit dug in the ground to hold drilling fluid or waste materials discarded after the treatment of drilling mud. Today, a mud pit is a steel rectangular tank, usually open at the top, in which drilling fluid is placed on the rig. Several mud pits are used and are named according to their use in the circulating system. For example, the mud pump takes in mud from a suction pit, sediments in the mud fall out in a settling pit, and mud is stored in a storage pit. Although mud pits are steel tanks, they are often referred to as pits. However, "mud tanks" is the preferred terminology.

mud pit volume measuring device *n*: an instrument that transmits either a pneumatic or electrical signal from the mud pits to recorders and alarm devices at the driller's station. It monitors the level of mud in the pits and detects fluid gain or loss.

mud saver valve *n*: a lower kelly cock. See *drill stem safety valve*, *lower kelly cock*.

mud seal *n*: a synthetic rubber, ring-shaped washer that fits between parts of a device that are exposed to drilling mud and parts that need to be protected from drilling mud.

mud weight *n*: a measure of the density of a drilling fluid often expressed as a weight per unit volume, such as pounds per gallon, pounds per cubic foot, or kilograms per cubic metre. Mud weight is directly related to the amount of pressure the column of drilling mud exerts at the bottom of the hole.

multiplex electronic control (MUX) system *n*: an arrangement of equipment employed on floating rigs drilling in water depths greater than 5,000 feet (1,500 metres) to overcome delays in the transmission of signals to close and open the subsea blowout preventers. Surface electronics transmit electronic command signals through cable reels to subsea multiplex electronics packages, which decode and deliver the commands to solenoid valves.

multiposition lock *n*: on Hydril ram blowout preventers, a hydraulically operated device that automatically secures (locks) the rams closed after they are closed normally by the blowout preventer control unit (accumulator). It consists of two clutch plates with serrated teeth that allow the rams to be opened only when hydraulic pressure is applied to separate the plates.

MUX *abbr*: multiplex electronic control.

N

neoprene *n*: a synthetic rubber made by the polymerization of chloroprene; it resists being broken down when exposed to various chemicals and oil. Neoprene can be used for an annular preventer's packing element when oil-based muds are being used at temperatures ranging from –30°F to 170° F.

nitrile rubber *n*: a synthetic rubber formed by polymerization of acrylonitrile with butadiene. Nitrile rubber can be used as a packing element in an annular preventer where oil-based mud is being used at temperatures ranging from 30°F to 180°F.

nonrising stem *n*: on a gate valve, the steel threaded rod (the stem) that, when rotated by an operating wheel, moves the valve's opening up or down in the valve body to open or close the valve, but does not move into the valve body. A nonrising stem prevents increases in valve body pressure that would occur if the stem entered the body.

O

open hole *n*: 1. any wellbore in which casing has not been set. 2. open or cased hole in which no drill pipe or tubing is suspended. 3. the portion of the wellbore that has no casing.

operating pressure *n*: in a blowout preventer stack, the amount of hydraulic pressure supplied by the blowout preventer control unit that is required to open and to close a blowout preventer. See *blowout preventer control unit*.

orifice *n*: 1. an opening, mouth, or a vent. 2. an open space allowing passage, such as an aperture, hole, mouth, opening, outlet, or vent.

outer barrel *n*: in the telescopic joint of a marine-riser assembly, a pipe that attaches to the top joint of the marine-riser assembly and to which are attached the riser tensioner lines. The inner barrel of the telescopic joint moves up and down within the outer barrel as the floating offshore rig heaves.

overboard vent line *n*: in a diverter system, a pipe (a line) that is permanently attached to side outlets on the diverter's housing. Usually, two vent lines are provided so that well fluids can be safely vented away from the rig, regardless of the wind's direction. See *diverter*.

P

packing element *n*: in an annular blowout preventer, the rubber or synthetic rubber doughnut-shaped piece that, when actuated, closes around pipe, wireline, or on open hole to seal the annulus and prevent the escape of well fluids to the rig floor.

pack-off nut *n*: in a subsea wellhead assembly, a device that, when rotated, actuates a seal assembly to close the annulus between two casing strings.

padeye *n*: a steel plate with an opening (the eye) that is usually welded to a heavy object and to which a sling or other lifting line is attached to facilitate the object's being moved, lifted, or lowered.

permanent guide base *n*: a structure attached to and installed with the foundation pile when a well is drilled from an offshore floating drilling rig. It is seated in the temporary guide base and serves as a wellhead housing. Guidelines are attached to it so that equipment, such as the blowout preventers, may be guided into place on the wellhead.

pilot manifold *n*: a device on a floating offshore drilling rig's accumulator that receives regulated pressure of 2,000 psi (14,000 kPa) and that contains several ¼-in. (6.35-mm) pilot control valves. The control valves send a pilot signal to the subsea control pods to operate the subsea blowout preventers.

pilot regulator *n*: a special valve (a regulator) on a floating rig's accumulator that sends variable pressure pilot signals, from 0 to 3,000 psi (21,000 kPa), to regulators mounted on the subsea control pod. A pilot regulator is pneumatically operated and has a 30 to 1 ratio, which means that the hydraulic output pressure is 30 times the air pilot signal.

pin connector *n*: in floating offshore drilling operations, a tool that connects the marine riser to the subsea casinghead on the seafloor.

pipe ram *n*: a sealing component for a blowout preventer that closes the annular space between the pipe and the blowout preventer or wellhead.

PLC *abbr*: programmable logic controller.

pod latch mechanism *n*: in a marine-riser system on floating offshore rigs, a device on a retrievable control pod that allows the pod to be remotely connected and disconnected from the lower marine-riser package's baseplate.

pod regulator *n*: a special valve (a regulator) that reduces and maintains (regulates) accumlator supply pressure to a subsea blowout preventer's normal working pressure. Two pod regulators exist, one designated as the BOP manifold regulator, and one designated as the annular regulator.

pod-select valve *n*: on a floating offshore drilling rig using a subsea blowout preventer stack and marine-riser system, a control device (a valve) that distributes accumulator fluid from the surface to a selected subsea control pod located on the lower marine-riser package.

Poisson's effect *n*: movement that results from structural compression caused by pressure exerted on the ends of the male (pin) portion of a box-and-pin connection.

preloading *n*: the tightening of blowout preventer (BOP) components to final torque under pressure high enough to simulate well pressures. Preloading prevents separation of the components from well pressures and, in the case of subsea BOP stacks, from bending forces caused by currents and rig movements.

pressure gauge *n*: an instrument that measures fluid pressure and usually registers the difference between atmospheric pressure and the pressure of the fluid by indicating the effect of such pressures on a measuring element (e.g., a column of liquid, pressure in a Bourdon tube, a weighted piston, or a diaphragm).

pressure gradient *n*: a scale of pressure differences in which there is a uniform variation of pressure from point to point. For example, the pressure gradient of a column of water is about 0.433 pounds per square inch per foot (9.794 kilopascals per metre) of vertical elevation. The normal pressure gradient in a formation is equivalent to the pressure exerted at any given depth by a column of 10 percent salt water extending from that depth to the surface (0.465 pounds per square inch per foot, or 10.518 kilopascals per metre).

pressure rating *n*: the operating (allowable) internal pressure of a vessel, tank, or piping used to hold or transport liquids or gases.

pressure switch *n*: on an accumulator, a control housed in an explosion proof junction box that is energized by hydraulic pressure. Hydraulic pressure turns the switch on or off.

pressure transducer *n*: an electronic device that senses fluid pressure and converts the pressure to electrical voltage. This voltage, in turn, can actuate other devices, such as meters on remote panels, which provide pressure readouts of all system pressures. See *transducer*.

production casing *n*: the last string of casing set in a well, inside of which is usually suspended a tubing string.

production liner *n*: a liner that functions as production casing. See *liner*, *production casing*.

programmable logic controller (PLC) *n*: a device used to manage, or control, another device or devices that govern the operation of a system or process. An operator, using an attached computer, can program the controller to maintain a given set of desirable circumstances and to respond to changes or upsets in the system or process using ladder logic, which is a logic system that operates much like the rungs on a ladder—that is, before the next rung on the ladder can be scaled, the controller must determine that certain conditions are met on the current rung.

pup joint *n*: a length of drill or line pipe, tubing, or casing shorter than range 1 (18 feet or 6.26 metres for drill pipe) in length.

R

radial packer seal *n*: on the telescopic joint of a marine-riser system, a flexible, strong rubber ring between the joint's inner and outer barrel that keeps mud in the inner barrel from entering the outer barrel.

ram *n*: the closing and sealing component on a blowout preventer. One of three types—blind, pipe, or shear—may be installed in several preventers mounted in a stack on top of the wellbore. Blind rams, when closed, form a seal on a hole that has no drill pipe in it; pipe rams, when closed, seal around the pipe; shear rams cut through drill pipe and then form a seal.

ram block *n*: see *ram*.

ram blowout preventer *n*: a blowout preventer that uses rams to seal off pressure on a hole that is with or without pipe. Also called a ram preventer. Compare *annular blowout preventer.*

ram bonnet *n*: the housing on a ram blowout preventer inside of which the rams and ram operating parts move when the preventer is operated.

ram cavity *n*: see *ram bonnet.*

ram packer *n*: the sealing device in a ram blowout preventer that is usually made of synthetic rubber. The packer is molded to the ram's upper and lower steel antiextrusion plates.

ram preventer *n*: see *ram blowout preventer.*

redundancy *n*: any deliberate duplication or partial duplication of circuitry or information to decrease the probability of a system or communication failure. In the transmission of information, the fraction of the gross information content of a message which can be eliminated without loss of essential information.

reeve *v*: to pass (as a rope) through a hole or an opening; to pass a rope over or through sheaves (pulleys) in a block.

regulator *n*: a device that reduces the pressure or volume of a fluid flowing in a line and maintains the pressure or volume at a specified level.

relative motion *n*: in a marine-riser system, movement caused by the structural compression that occurs when pressure is exerted on the ends of the pins in the riser connectors (Poisson's effect), temperature differences between the fluid in the main riser and the fluids in the auxiliary lines, and bending loads imposed by deflections of the riser. Such motion of the riser connectors can cause fatigue cracking of the support flange if an adequate gap is not provided between the support flange and the coupling. See *Poisson's effect.*

remotely operated vehicle (ROV) *n*: in offshore operations, an underwater device controlled from a vessel on the water's surface that is used to inspect and operate certain devices on subsea equipment, such as a blowout preventer stack, and that can be used in place of or in conjunction with diving personnel.

reservoir *n*: a container or vessel that stores fluid, as a reservoir on an accumulator that holds hydraulic operating fluid.

RF flanged connector *n*: a brand of hydraulic connector used in subsea marine-riser systems. It is similar to Cameron's Load King connector but has a tensile load rating of 2 million pounds (900 tonnes). See *Load King connector.*

rig *n*: the derrick or mast, drawworks, and attendant surface equipment of a drilling or workover unit.

rig floor *n*: the area immediately around the rotary table and extending to each corner of the derrick or mast—that is, the area immediately above the substructure on which the rotary table and other equipment rest. Also called the derrick floor, drill floor.

riser *n*: a pipe through which liquid travels upward. See *riser pipe.*

riser adapter *n*: a fitting installed on top of the lower marine-riser package's (LMRP's) flex joint. It is a fitting to which the first riser joint is attached to the LMRP.

riser assembly *n*: see *riser pipe*.

riser collapse *n*: the caving in of a riser joint or several riser joints that can occur if the pressure inside the riser drops below external seawater pressure.

riser collapse resistance *n*: the ability of the riser pipe to withstand its tendency to cave in if pressure inside the riser drops below external seawater pressure.

riser connector *n*: a fitting attached to the end of each riser joint that allows the joints to be attached to one another.

riser flange *n*: on a riser connecter, the flat, ring-shaped fitting that provides a place for riser joints to be attached to each other.

riser handling tool *n*: a special tool that enables riser joints, as well as the telescopic joint, to be raised and lowered into and out of the water. See *hydraulic riser handling tool, mechanical riser handling tool.*

riser joint *n*: see *riser pipe*.

riser joint connector *n*: see *riser connector*.

riser pipe *n*: the pipe and special fittings used on floating offshore drilling rigs to establish a seal between the top of the wellbore, which is on the ocean floor, and the drilling equipment, located above the surface of the water. A riser pipe serves as a guide for the drill stem from the drilling vessel to the wellhead and as a conductor of drilling fluid from the well to the vessel. The riser consists of several sections of pipe and includes special devices to compensate for any movement of the drilling rig caused by waves. Also called marine-riser pipe, riser joint.

riser pup joint *n*: a short length—usually, from 5 to 40 feet (1 to 12.2 metres) long—of riser pipe that is used to space out the riser assembly as it is run to various water depths. See *space-out*.

riser spider assembly *n*: see *riser support spider*.

riser string *n*: see *riser pipe*.

riser support gimbal *n*: in deepwater applications on floating rigs, a device placed between the rotary table and riser support spider to compensate for offset caused by ocean movements. Hydraulic cylinders, or flex elements, support the weight of the spider, the entire riser string, and BOP assembly. It also cushions shock loads imparted on the system. See *riser support spider*.

riser support spider *n*: a fitting placed in the rotary table of an offshore floating rig, which supports a joint of riser pipe as the pipe is being lowered into or pulled from the water. The spider consists of a base plate and top plate with a series of support dogs, or arms, between them. The dogs support the riser joint under the riser joint's support flange. Hydraulic cylinders force the support dogs inward and under the riser joint support flange. They are also retracted hydraulically to allow the riser joints to pass through. On some riser spiders, the support arms are engaged and disengaged manually. The riser and BOP assembly rest on the riser support spider as the crew picks up and adds each new joint to the string. The riser spider's ID is the same as the rotary table's ID, and the base plate is designed for the specific rotary table with which it is used.

riser system *n*: see *riser pipe*.

riser tensioner *n*: an assembly of strong cables (lines) connected to the outer barrel of the telescopic joint and a hydropneumatic piston-and-cylinder sheave assembly. The tensioning system supports the weight of the riser joints by applying force (tension) to the outer barrel of the telescopic joint. Tensioner lines are connected to a tensioner ring or to fixed padeyes on the outer barrel. The applied tension must remain constant while compensating for vessel heave.

riser tensioner line *n*: a cable that supports the marine riser while compensating for vessel movement. See *riser tensioner*.

rotary hose *n*: a steel-reinforced, flexible hose that is installed between the standpipe and the swivel or top drive. It conducts drilling mud from the standpipe to the swivel or top drive. Also called the kelly hose or the mud hose.

rotary table *n*: the principal piece of equipment in the rotary table assembly; a turning device used to impart rotational power to the drill stem while permitting vertical movement of the pipe for rotary drilling. The master bushing fits inside the opening of the rotary table; it turns the kelly bushing, which permits vertical movement of the kelly while the stem is turning.

ROV *abbr*: remotely operated vehicle.

ROV intervention system *n*: in a subsea blowout prevention system with a marine riser, the equipment that allows a remotely operated vehicle (ROV) to tie into and operate certain functions, such as closing the shear rams and unlocking the lower marine-riser package (LMRP) when the normal surface operating system fails. A ported hydraulic stabbing device in the ROV mates with a receptacle in the LMRP's hydraulic circuit and pumps hydraulic fluid into the circuit to operate the selected component.

running tool *n*: any specialized tool used to run equipment in a well, such as a marine-riser joint or a wireline running tool for installing retrievable gas-lift valves. Various tubing-type running tools are also used.

S

safety valve *n*: a control device (a valve) installed at the top of the drill stem, which, when closed, prevents flow out of the drill pipe if a kick occurs during tripping operations.

seabed *n*: see *seafloor*.

seafloor *n*: the bottom of the ocean; the seabed.

seal assembly *n*: apparatus that seals the annulus between each casing string, and isolates the previous casing from the higher wellbore pressures encountered at greater depths.

secondary circuit *n*: in a riser connector, a way of routing hydraulic fluid to provide additional unlocking force that may be required to overcome high friction forces generated between the tapers on the cam ring and the locking dogs, or segments.

SEM *abbr*: subsea electronic module.

separation *n*: the process of removing gas from liquid and liquid from gas.

shear *n*: action or stress that results from applied forces and that causes or tends to cause two adjoining portions of a substance or body to slide relative to each other in a direction parallel to their plane of contact.

shear-blind ram *n*: a blind-shear ram. See *shear ram*.

shear ram *n*: the component in a blowout preventer that cuts, or shears, through drill pipe and forms a seal against well pressure. Shear rams are used in floating offshore drilling operations to provide a quick method of moving the rig away from the hole when there is no time to trip the drill stem out of the hole.

shear ram preventer *n*: a blowout preventer that uses shear rams as closing elements. See *shear ram*.

sheave *n*: (pronounced "shiv") 1. a grooved pulley. 2. support wheel over which tape, wire, or cable rides.

shut-in *adj*: shut off to prevent flow. Said of a well, plant, pump, and so forth, when valves are closed at both inlet and outlet.

shut in *v*: to close in a well in which a kick has occurred.

shuttle valve *n*: a flow control device (a valve) with two inlet ports and one outlet port in an accumulator, which is employed in a subsea blowout preventer system. Hoses from the lower marine-riser package's yellow and blue control pods are connected to the shuttle valve's inlet port. The outlet port is connected directly to the blowout preventer's inlet port.

side entry *n*: an opening in the wall of the riser adapter to which a mud booster line is attached. See *mud booster line*, *riser adapter*.

side outlet *n*: an opening usually located in the blowout preventer stack below a ram preventer, to which a choke line or a kill line is attached. Typically, outlets are provided in pairs, one on each side of the bore. Usually, outlets are provided for a choke line and a kill line.

SI system *n*: see *International System of Units*.

sleeve *n*: a tubular part designed to fit over another part.

slick riser joint *n*: in drilling from deepwater floating offshore drilling rigs, a riser joint that does not have buoyant riser modules attached to it. See *buoyant riser joint*, *riser pipe*.

slip joint *n*: see *telescopic joint*.

slips *n pl*: wedge-shaped pieces of metal with serrated inserts (dies) or other gripping elements, such as serrated buttons, that suspend the drill pipe or drill collars in the master bushing of the rotary table when it is necessary to disconnect the drill stem from the kelly or from the top-drive unit's drive shaft. Rotary slips fit around the drill pipe and wedge against the master bushing to support the drill collar. Power slips are pneumatically or hydraulically actuated devices that allow the crew to dispense with the manual handling of slips when making a connection.

solenoid valve *n*: in an accumulator used on a floating offshore drilling rig, a flow control device (a valve) that, when energized by operating a blowout preventer, opens to allow rig air pressure to flow to a pneumatic cylinder, which, in turn, operates the blowout preventer.

space-out *n*: 1. the installation of one or more riser pup joints to obtain the overall required length of the riser from the blowout preventer stack to the telescopic joint. 2. the act of ensuring that a pipe ram preventer will not close on a drill pipe tool joint when the drill stem is stationary. A pup joint may be made up in the drill string to lengthen it sufficiently.

spider *n*: see *riser support spider.*

SPM *abbr*: subsea pilot manipulated.

spring-loaded lock *n*: in a dog-type riser connector, a device that prevents the actuator screws in the connector from backing out because of vibration. See *dog-type riser connector*.

spring-loaded pin *n*: in an ABB Vetco Gray riser connector, one of several pins that protrude from the connector's body and fit under a support shoulder on the gasket. Hydraulic force releases the pins and spring force retains the gasket. See *riser connector*.

stab *n*: a type of connector on tools in which the joining, or mating, of the tools is achieved by inserting a pin on one tool into a receptacle on the other. *v*: to guide and insert the pin of one tool into the receptacle of another, as the pin on a drill pipe's tool joint into the box of another tool joint when making a connection.

standard riser joint *n*: a joint of riser pipe that is of normal length—that is, it is not a riser pup joint. See *riser pipe*, *riser pup joint*.

standpipe *n*: a vertical pipe rising along the side of the derrick or mast, which joins the discharge line leading from the mud pump to the rotary hose and through which mud is pumped into the hose.

standpipe manifold *n*: a part of the mud circulation system that contains several valves, piping, tees, and elbows and is part of the standpipe. Manifold valves direct the mud from the mud pumping system through the piping and fittings and allow the rig crew to direct the mud to its desired destination, such as to the rotary hose, a backup standpipe, or other piece of circulating equipment. See *standpipe*.

static load *n*: a nonvarying weight; the force exerted by the weight of a mass at rest.

string *n*: an assembly of several individual joints, or lengths, of tubulars, such as drill pipe, drill collars, casing, and tubing or other lengths of equipment, such as sucker rods. The tubulars or rods are joined to form a continuous length of pipe or rods. For example, connecting several single joints of drill pipe form a string of drill pipe that can be thousands of feet or metres long; or connecting several lengths of sucker rods form a string of sucker rods that can also be thousands of feet or metres long.

stroke length *n*: in a telescopic joint, the distance the inner barrel moves within the outer barrel. It can be from 45 to 65 feet (14 to 20 metres) depending on the rig requirements.

structural casing *n*: see *conductor casing*.

subsea accumulator *n*: a device mounted on or near a subsea blowout preventer stack, which stores hydraulic operating fluid and, because of its close proximity to the stack, shortens the length of time required to close a preventer. It may also store fluid under high pressure to close a shear ram.

subsea blowout preventer *n*: a blowout preventer placed on the seafloor for use by a floating offshore drilling rig. See *blowout preventer*.

subsea completion *n*: a well completion in which the flow of hydrocarbons from the well is controlled by equipment placed on or below the seafloor. See *well completion*.

subsea control pod *n*: see *hydraulic control pod*.

subsea electronic module (SEM) *n*: in drilling from floating drilling rigs in deep water, a device in a multiplex electronic (MUX) control system that contains a programmable logic controller, a modem, and a power supply. A SEM converts signals received from the blowout preventer operating system on the surface to ensure quick operation of the subsea blowout preventer components. See *multiplex electronic (MUX) control system*.

subsea pilot manipulated (SPM) *adj*: of a device that is operated and controlled (manipulated) by a subsea pilot valve. See *subsea pilot manipulated (SPM) valve*.

subsea pilot manipulated (SPM) valve *n*: on the control pod of a subsea blowout preventer system, a control device (a valve) employed in the operation of the blowout preventers. An SPM valve is a hydraulically operated, three-position, four-way valve that directs regulated pressure to the blowout preventers.

subsea wellhead *n*: the equipment installed at the top of the wellbore, which is below the water's surface and on the seabed, and to which the blowout preventer stack is attached.

subsea wellhead and casing system *n*: in offshore drilling from floating rigs, the equipment and pipe (casing) that forms the foundation of a well and links the wellbore to the subsea blowout preventer stack.

support flange *n*: on a riser joint of a floating offshore drilling rig, the flat area at each end of the joint to which other joints are connected. The flange supports static and dynamic loads on the riser system, as well as the auxiliary lines and buoyancy modules, if modules are fitted. See *riser pipe*.

support gimbal *n*: in deepwater offshore drilling operations from a floating rig, a device between the rotary table and the riser support spider. The support gimbal has hydraulic cylinders, or flex elements, that support the spider, the riser, and the blowout preventer stack. It compensates for riser offset caused by water motions, yet maintains the riser flange square to the drill floor to facilitate the making of connections. It also cushions shock loads imparted on the system. See *riser support spider*.

surface accumulator unit *n*: a storage and operating device on a floating offshore drilling rig that contains hydraulic fluid reservoirs, a fluid-mixing system, pumps, accumulator storage bottles, flowmeter, pod select valve, pilot manifold, hydraulic junction boxes, pressure gauges, and electrical junction boxes required to operate the subsea blowout preventer components. See *accumulator*.

surface casing *n*: see *surface pipe*.

surface pipe *n*: the first string of casing (after the conductor pipe) that is set in a well. It varies in length from a few hundred to several thousand feet (metres). Some states require a minimum length to protect freshwater sands. Also called surface casing.

surging *n*: a rapid increase in pressure downhole that occurs when the drill stem is lowered too fast or when the mud pump is brought up to speed after starting.

swab *v*: to pull formation fluids into a wellbore by raising the drill stem at a rate that reduces the hydrostatic pressure of the drilling mud below the bit.

swabbing *n*: see *swabbing effect*.

swabbing effect *n*: a phenomenon characterized by formation fluids being pulled, or swabbed, into the wellbore when the drill stem and bit are pulled up the wellbore fast enough to reduce the hydrostatic pressure of the mud below the bit. If enough formation fluid is swabbed into the hole, a kick can result.

swivel *n*: a rotary tool that is hung from the hook and the traveling block to suspend and permit free rotation of the drill stem. It also provides a connection for the rotary hose and a passageway for the flow of drilling fluid into the drill stem.

swivel tensioner ring *n*: in drilling from floating offshore drilling rigs, a device installed on top of a riser joint to which tensioner lines are manually installed to padeyes on the ring after the joint is lowered through the rotary table. Tensioner lines must be manually removed from the ring when a riser joint is being raised through the rotary table. See *riser pipe*, *riser tensioner*.

syntactic foam *n*: a material used in the manufacture of buoyant riser modules that contains numerous spheres with hard shells that trap air. The spheres are embedded in a durable foam-like material that is molded into a shape that will fit around a slick riser joint (a joint of riser pipe that does not have buoyant modules added to it). See *buoyant riser joint*, *riser pipe*.

syntactic foam module *n*: see *buoyant riser joint*.

T

tee *n*: a pipe fitting that is shaped like the letter T. A pipe or other fitting can be attached to each end of the tee, since the tee ends are threaded.

telescopic joint *n*: a device used in the marine-riser system of a floating drilling rig to compensate for the vertical motion of the rig caused by wind, waves, or weather. It consists of an inner barrel attached beneath the rig floor and an outer barrel attached to the riser pipe and is an integrated part of the riser system. Also called slip joint.

temporary guide base *n*: the initial piece of equipment lowered to the ocean floor once a floating offshore drilling rig has been positioned on location. It serves as an anchor for the guidelines and as a foundation for the permanent guide base and has an opening in the center through which the bit passes. It is also called a drilling template.

tensile load *n*: the amount of longitudinal stress borne by an object or substance.

tensioner system *n*: a set of devices installed on a floating offshore drilling rig to maintain constant tension on the riser pipe, despite any vertical motion made by the rig.

tensioner unit *n*: part of the riser tensioner system on a floating offshore drilling rig that uses a hydropneumatic cylinder assembly to which is attached the tensioning lines from the telescopic joint. High-pressure air and hydraulic fluid in the cylinder assembly provide tension on the lines as the rig moves up and down (heaves) with ocean movements.

termination joint *n*: see *termination spool.*

termination spool *n*: a standard riser joint with a side entry to allow connection of a mud booster line. It is used to overcome a height restriction, and is the first riser joint connected to the lower marine-riser package (LMRP). Some spools feature an automatic check valve that allows mud to flow only down the circulation line and up the riser string.

test cap *n*: on choke and kill lines that are attached to a riser joint, a pressure-tight stopper (a cap) fitted on the top of the choke and kill lines. When the lines are capped, crew members can apply pressure to them and ensure that they do not leak as they run the riser pipe.

test plug *n*: when pressure testing a blowout preventer stack, a blocking device that is run and landed in the high-pressure wellhead housing. The plug isolates the casing string, which has a lower pressure rating than the blowout preventer stack, from blowout preventer test pressure.

thread *n*: a continuous helical rib, as on a screw or pipe.

top drive *n*: a device similar to a power swivel that is used in place of the rotary table to turn the drill stem. It also includes power tongs. Modern top drives combine the elevator, the tongs, the swivel, and the hook. Even though the rotary table assembly is not used to rotate the drill stem and bit, the top-drive system retains it to provide a place to set the slips to suspend the drill stem when drilling stops.

top receiver plate *n*: a flat steel piece (a plate) that houses the choke and kill line connections to the lower marine-riser package (LMRP), the blowout preventer's control system receptacles, and the LMRP's guidance devices.

top seal *n*: in a ram blowout preventer, a piece of synthetic rubber that is molded to the upper and lower steel antiextrusion plates of the ram block.

torque *n*: the turning force that is applied to a shaft or other rotary mechanism to cause it to rotate or tend to do so. Torque is measured in foot-pounds, joules, newton-metres, and so forth.

transducer *n*: a substance or a device, such as a piezoelectric crystal, a microphone, or a photoelectric cell, that converts input energy of one form into output energy of another. For example, a telephone receiver receives electric power and supplies acoustic power.

trip tank *n*: a small mud tank with a capacity of 10 to 15 barrels (1 to 3 cubic metres), often with 1-barrel or ½-barrel (decalitre or litre) divisions, used to ascertain the amount of mud necessary to keep the wellbore full with the exact amount of mud that is displaced by drill pipe. When the bit comes out of the hole, a volume of mud equal to that which the drill pipe occupied while in the hole must be pumped into the hole to replace the pipe. When the bit goes back in the hole, the drill pipe displaces a certain amount of mud, and a trip tank can be used again to keep track of this volume.

true vertical depth (TVD) *n*: the depth of a well measured from the surface straight down to the bottom of the well. The true vertical depth of a well may be quite different from its actual measured depth, because wells are very seldom drilled exactly vertical.

U

UEPS *abbr*: uninterruptible electrical power supply.

uncontrolled blowout *n*: see *blowout*.

uninterruptible electrical power supply (UEPS) *n*: an electrical system with built-in redundancy that supplies (powers) an entire electronic control system, and provides immediate changeover should one part of the system fail.

upper kelly cock *n*: a valve installed above the kelly that can be closed manually to protect the rotary hose from high pressure that may exist in the drill stem.

V

VAC *abbr*: volts, alternating current.

vacuum degasser *n*: a device in which entrained gas in the mud returning from the wellbore is removed from the mud by the action of a vacuum inside a tank. The gas-cut mud is pulled into the tank, the gas removed, and the gas-free mud discharged back into the mud pits.

variable bore ram *n*: a ram blowout preventer that contains blocks (rams) that can close and seal on a range of pipe sizes—for example, from 3½- to 5-inch (89- to 127-millimetre) pipe. Variable bore rams contain a large reserve of rubber in the ram block that specially designed antiextrusion plates force into sealing contact with smaller sizes of pipe. The antiextrusion plates also support the excess rubber when wellbore pressure is applied.

VDC *abbr*: volts, direct current.

vessel offset *n*: a phenomenon that occurs in drilling from floating offshore rigs (vessels) in which the vessel on the surface is not located directly above the wellhead on the seafloor. While anchoring and dynamic positioning techniques hold offset to a minimum, a certain amount occurs and subsea equipment must be designed to compensate for offset.

VIV *abbr*: vortex-induced vibration.

volts, alternating current (VAC) *n*: electric current that flows through an electrical circuit in an alternating manner—that is, the current reverses direction in the circuit at regular intervals.

volts, direct current (VDC) *n*: electric current that flows through an electrical circuit in one direction only.

vortex *n*: a whirling mass, e.g., a whirlpool.

vortex-induced vibration (VIV) *n*: a shaking motion (a vibration) created in a marine-riser system by the tendency of water to swirl (create a vortex) around the riser.

W

wear bushing *n*: a fixed or removable cylindrical metal lining placed inside a fitting or piece of moving equipment that reduces friction and thus wear on the fitting.

wear sleeve *n*: in a riser adapter, a thin metal cylinder that fits inside the adapter and prevents premature wear on the inside bore of the adapter and drill stem components that rotate within it. See *riser adapter*.

Wedge Lock *n*: a device on Cameron ram blowout preventers that locks the rams on a subsea preventer closed so that even if hydraulic closing pressure is lost, the rams will remain closed and provide a seal. It is a wedge-shaped piston that is mounted in its own hydraulic housing on the back of each ram bonnet at 90 degrees to the operating piston's tail shaft. Once the rams are closed, the Wedge Lock pistons are hydraulically operated and they move behind the operating piston tail shaft and wedge it into place.

well *n*: the hole made by the drilling bit, which can be opened, cased, or both. Also called borehole, hole, or wellbore.

wellbore *n*: a borehole; the hole drilled by the bit. A wellbore may have casing in it or it may be open (uncased); or part of it may be cased, and part of it may be open. Also called a borehole or hole.

well completion *n*: 1. the activities and methods of preparing a well for the production of oil and gas or for other purposes, such as injection; the method by which one or more flow paths for hydrocarbons are established between the reservoir and the surface. 2. the system of tubulars, packers, and other tools installed beneath the wellhead in the production casing; that is, the tool assembly that provides the hydrocarbon flow path or paths.

well control *n*: the methods used to control a kick and prevent a well from blowing out. Such techniques include, but are not limited to, keeping the wellbore completely filled with drilling mud of the proper weight, or density, during all operations, exercising reasonable care when tripping pipe out of the hole to prevent swabbing, and keeping careful track of the amount of mud put into the hole to replace the volume of pipe removed from the hole during a trip.

well-control equipment *n*: an assembly of several components, such as ram preventers, annular preventers, a choke and kill system, trip tanks, and mud-gas separators. On offshore floating rigs, well-control equipment also includes a marine-riser system and a diverter system.

well fluid *n*: the fluid, usually a combination of gas, oil, water, and suspended sediment, that comes out of a reservoir. Also called well stream.

wellhead *n*: the equipment installed at the top of the wellbore. A wellhead includes such equipment as the casinghead and tubing head. *adj*: pertaining to the wellhead (e.g., wellhead pressure).

wellhead connector *n*: on a floating drilling rig using a subsea blowout preventer stack, a tool that attaches the blowout preventer (BOP) stack to the wellhead. It is hydraulically operated from the surface and, when actuated, connects the BOP to the wellhead.

well kick *n*: see *kick*.

wireline *n*: a slender, rodlike or threadlike piece of metal usually small in diameter, that is used for lowering special tools (such as logging sondes, perforating guns, and so forth) into the well. Also called slick lines. Compare *wire rope.*

wire rope *n*: a cable composed of steel wires twisted around a central core of fiber or steel wire to create a rope of great strength and considerable flexibility. Wire rope is used as drilling line (in rotary and cable-tool rigs), coring line, servicing line, winch line, and so on. It is often called cable or wireline; however, wireline is a single, slender metal rod, usually very flexible. Compare *wireline.*

working pressure *n*: the maximum pressure at which an item is to be used at a specified temperature.

Y

yield strength *n*: in a joint of riser pipe, a measure of the amount of pressure inside or outside the riser that is required to permanently distort it.

Review Questions

LESSONS IN ROTARY DRILLING

Unit V, Lesson 10: Marine Riser Systems and Subsea Blowout Preventers

True or False

Put a T for *true* or an F for *false* in the blank next to each statement.

_________ 1. When the mud weight is measured in pounds per gallon (ppg), the constant used to calculate hydrostatic pressure in pounds per square inch (psi) is 0.053.

_________ 2. A well is 10,861 feet (ft) deep and is full of a mud that weighs 11.6 ppg. The hydrostatic pressure that mud exerts at the bottom of the well is 6,551 psi.

_________ 3. A well kick is the entry of formation fluids into the wellbore.

_________ 4. Because gas-cut mud is caused by the entry of relatively small volumes of gas into the wellbore, it is usually not necessary to degas the mud.

_________ 5. Pulling the drill stem from the hole too fast can cause swabbing.

_________ 6. Abnormal pressure plays no role in causing kicks.

_________ 7. A blowout is the entry of formation fluids into the wellbore.

_________ 8. If rig crew members fail to keep the hole full while removing pipe from the hole, a kick can occur.

_________ 9. The first line of defense against kicks is the blowout preventer equipment.

_________ 10. Floating rigs usually employ subsea blowout prevention equipment, while bottom-supported units, such as jackups, do not.

Multiple Choice

Pick the *best* answer from the choices and place the letter of that answer in the blank provided.

_________ 11. A marine riser—

a. provides a fluid conduit to and from the wellbore.
b. supports auxiliary lines.
c. guides the drill stem and other tools from the drilling rig to the wellhead on the seabed.
d. provides a means of running and retrieving the BOP assembly from the surface to the wellhead on the seafloor.
e. all of the above

_________ 12. A riser joint with an outside diameter (OD) of 21 inches (in.) or 533.4 millimetres (mm) is commonly used on a blowout preventer (BOP) with an inside diameter (ID) of—
   a. 13⅝ in. (346 mm).
   b. 16¾ in. (425.5 mm).
   c. 18¾ in. (476.3 mm).
   d. 21¾ in. (539.8 mm).

_________ 13. Riser pup joints—
   a. are the same length as normal riser joints.
   b. range in length from 5 to 40 ft or 1.5 to 12.2 metres (m).
   c. are available in lengths of 40 ft (12.2 m) only.
   d. bark somewhat like a puppy when lowered into the water.

_________ 14. An automatic flooding valve—
   a. prevents riser collapse.
   b. when open, allows seawater to fill the riser.
   c. can be operated manually if required.
   d. all of the above

_________ 15. Two main types of riser connector are the—
   a. dog-type and bolted-flange type.
   b. cat-type and screwed-flange type.
   c. cat- and dog-type and pressure-flange type.
   d. none of the above

_________ 16. Where height restrictions occur, a device that can be used to connect the first riser joint to the top of the flex joint is the—
   a. Load King connector.
   b. termination spool, or joint.
   c. LMRP.
   d. RF connector.

_________ 17. A telescopic joint—
   a. compensates for vertical movement (heave) of the vessel.
   b. provides a means of connecting the diverter assembly to the riser.
   c. provides terminations for the riser auxiliary lines to flexible hoses at the drilling vessel.
   d. provides attachment points for the riser tensioning system.
   e. all of the above

_________ 18. A hydraulic tensioner line allows crew members to—
   a. run and retrieve the telescopic joint without removing the tensioner lines.
   b. manually install the tensioner lines to the padeyes on the ring after the joint is lowered through the rotary.
   c. hydraulically latch the tensioner ring to the outer barrel as the joint is run.
   d. both a and c

_________ 19. Poisson's effect—
   a. is the tendency of the riser system to attract fish.
   b. ensures that a tight fit occurs on auxiliary lines where they join the riser system.
   c. creates motion between the box and pin on the auxiliary line couplings.
   d. is a temperature effect.

_________ 20. Choke and kill lines in a subsea BOP system allow the crew to circulate mud down—
   a. one line and take returns up the other.
   b. down the drill stem and take returns up one line.
   c. down the drill stem and take returns up both lines.
   d. all of the above
   e. none of the above

Fill in the Blanks

Fill in the blanks with an appropriate word or phrase. Pick the correct term from those listed below.

LMRP	riser-mounted conduit lines
ball joint	hydraulic riser handling tool
riser handling tools	flex joint
buoyant modules	mud booster line
flexible piping	mechanical riser handling tool

A 21. ____________________ is sometimes provided on riser joints to increase the return velocity of the fluid column in the riser. For deepwater applications, the BOP control system's hydraulic fluid is supplied through 22. ______________________________. The rig crew uses 23. ______________________________to raise and lower the riser and telescopic joints. A 24. _______________ duplicates a riser joint connection. A 25. _________________ ___is similar to a mechanical tool, but it is operated hydraulically to reduce the time involved in making the connections. Reducing the weight of the riser string is achieved by adding 26. ___________. A hydraulic connector allows the 27. ____________________ to be disconnected from the BOP stack. A 28. ___________________________ or a 29. ______________ _____________ is mounted between the annular preventer and the riser adapter on the LMRP. 30. ______________________ connects the choke and kill lines and the hydraulic conduit lines to the LMRP.

True or False

Put a T for *true* and an F for *false* in the blank next to each statement.

_________ 31. A riser adapter includes a side entry for the mud booster line, a check valve or gate valve for closing off the mud booster line, a riser connection, and the kick-out subs on the choke and kill lines and on the hydraulic conduit line.

_________ 32. A diverter system protects personnel and equipment by completely shutting in the well on a shallow flow.

_________ 33. The weight of the marine riser system must be supported while it is deployed.

_________ 34. In a riser tensioner system, it is not necessary to keep the air pressure on the system constant.

_________ 35. In water depths no greater than about 5,000 ft (1,500 m), guidelines are seldom used to guide the BOP and LMRP assembly to the wellhead on the seabed.

_________ 36. Equipment in the guideline tensioning system is basically the same as that found in the marine-riser tensioning system.

_________ 37. A high-pressure wellhead provides the foundation for the BOP stack, supports and houses subsequent casing strings, and provides a wellbore seal-and-locking arrangement between the surface casing and BOP stack.

_________ 38. On a floating rig, surface BOP stacks are usually installed.

_________ 39. It is not necessary for a floating rig's BOP stack to be able to hang off the drill string.

_________ 40. A guidelineless BOP stack frame is generally used in water depths exceeding 5,000 ft (1,500 m).

Matching

Write the letter of the correct definition in the blank next to each term.

Terms

_________ 41. subsea BOP stack

_________ 42. isolation test plug

_________ 43. ram blowout preventer

_________ 44. side outlet

_________ 45. choke and kill lines

_________ 46. choke manifold

_________ 47. stripping

_________ 48. hydraulic connector

_________ 49. blind-shear ram

_________ 50. pipe ram preventer

Definitions

a. Lowering the drill stem into the wellbore when the well is shut in on a kick and when the weight of the drill stem is sufficient to overcome the force of wellbore pressure.

b. An arrangement of piping and special valves, called chokes. In drilling, mud is circulated through it when the blowout preventers are closed.

c. Closes the annular space between drill pipe and the blowout preventer or wellhead.

d. Pipes, or conduits, usually installed on outlets below the ram blowout preventers, through which fluids in the annulus can flow to the surface and through the choke manifold when valves to it are open or that allows drilling mud or other fluid to be pumped into the well to control a well when it is not possible to pump down the drill string.

e. The component in a blowout preventer that cuts, or shears, through drill pipe and forms a seal against well pressure.

f. A special blocking device (a plug) that is used in pressure testing a blowout preventer. The plug is run and landed in a high-pressure wellhead housing, where it forms a pressure-tight seal. This seal protects (isolates) the casing below the wellhead from the high pressures used to test the blowout preventers, which are located above the wellhead and casing.

g. A blowout preventer placed on the seafloor for use by a floating offshore drilling rig.

h. A blowout preventer that uses rams to seal off pressure on a hole that is with or without pipe.

i. An opening usually located in the blowout preventer stack below a ram preventer, to which a choke line or a kill line is attached.

j. A device that attaches (connects) the blowout preventer stack to the wellhead and to the lower marine riser package. It is controlled from the surface.

Multiple Choicc

Pick the *best* answer from the choices and place the letter of that answer in the blank provided.

_________ 51. A blind-shear ram preventer—

a. closes in the well on drill pipe of a specific size.
b. can cut drill pipe and seal in the well.
c. is not often used in a subsea preventer stack.
d. should not be used except on shallow gas kicks.

________ 52. All of the following are true about operating and maintaining ram preventers *except*—
- a. Never inspect the ram packers and top seals.
- b. Install the correct size rams for the drill pipe in use.
- c. Do not apply pressure above the ram blocks.
- d. Accurately record the distance to each ram cavity from the rotary table.

________ 53. The maximum hydraulic operating pressure of ram preventers is—
- a. 3,000 psi (20,500 kPa).
- b. 1,500 psi (10,500 kPa).
- c. 750 psi (5,250 kPa).
- d. 500 psi (3,500 kPa).

________ 54. The normal hydraulic operating pressure of ram preventers is—
- a. 3,000 psi (20,500 kPa).
- b. 1,500 psi (10,500 kPa).
- c. 750 psi (5,250 kPa).
- d. 500 psi (3,500 kPa).

________ 55. Ram preventers have a way to lock the rams closed because—
- a. well pressure might open them.
- b. it is the only way to close the rams.
- c. it keeps the rams closed even if closing pressure is lost.
- d. none of the above

________ 56. The main function of an annular preventer is to—
- a. close in the well on drill pipe.
- b. close in the well on open hole.
- c. close and seal the wellbore and, at the same time, allow the drill stem to be moved through the closed preventer.
- d. close and seal the wellbore, but not allow the drill stem to be moved through the closed preventer.

________ 57. You are refurbishing an annular BOP and the driller tells you to install a new packer that is made of neoprene. Because of this requirement, you know that the well is probably going to be drilled using—
- a. water-base mud at temperatures ranging from –30 to 225°F (–34 to 107°C).
- b. oil-based mud at temperatures ranging from 30 to 180°F (–1 to 82°C).
- c. oil-based mud at temperatures ranging from –30 to 170°F (–34 to 77°C).

________ 58. Choke- and kill-line gate valves are usually operated—
- a. manually.
- b. hydraulically.
- c. neither manually nor hydraulically.
- d. none of the above

_________ 59. A balance stem below the gate of a gate valve—

a. keeps hydrostatic pressure created by seawater from overcoming spring force and opening the valve.
b. keeps hydrostatic pressure created by seawater from overcoming spring force and closing the valve.
c. balances the opening force of the valve against the hydrostatic pressure of seawater.
d. balances the closing force of the valve against the hydrostatic pressure of seawater.

_________ 60. A choke manifold is—

a. a valve that is adjusted to maintain the proper back-pressure on the well.
b. a series of pipes that chokes the well.
c. a series of gate valves and choke assemblies in a compact manifold that provides various flow paths for drilling fluids and well fluids encountered in well control.
d. installed on the subsea BOP to keep well pressures at a reasonable level.

True or False

Put a T for *true* or an F for *false* in the blank next to each statement.

_________ 61. A common type of choke assembly is a hydraulically operated, variable orifice valve that, when properly manipulated, controls back-pressure on the well as a kick is being circulated from the well.

_________ 62. A mud-gas separator separates the gas from the mud and vents the gas at a safe distance from the drill floor.

_________ 63. The pressure rating of a drill stem safety valve should be the same as the rig's ram BOPs.

_________ 64. An inside BOP is also called a Brown valve.

_________ 65. One advantage of a flapper-type float valve is that it permits the passage of balls and go-devils.

_________ 66. A trip tank is a large-volume tank that holds the enormous quantities of drilling mud needed to replace the drill stem as it is removed from the hole.

_________ 67. Upper and lower kelly cocks should have a pressure rating lower than the rig's ram BOP's pressure rating.

_________ 68. Normally, the standpipe's pressure rating is higher than the rig's ram preventers.

_________ 69. Mud pit volume measuring devices warn of an increase in the rate at which mud returns from the well.

_________ 70. A flow rate sensor monitors the level of mud in the pits and warns the driller of increases or decreases.

Fill in the Blanks

Fill in the blanks below with an appropriate word or phrase. Pick the correct term from those listed below.

50%	multiplex electronic control (MUX) systems
accumulator unit	hydraulic fluid
multiplexed	1,000 psi (7,000 kPa)
straight hydraulic	hose bundle
1,500 psi (10,500 kPa)	provides the pilot signal very quickly

The BOP control system must deliver, on command, 71. ______________________ at the correct pressure to operate the BOP components. Two types of indirect pilot system are the 72. ______________________ type and the 73. ______________________ type. A surface 74. ______________________ contains hydraulic fluid reservoirs, a fluid mixing system, pumps, accumulator storage bottles, a flowmeter, a pod select valve, a pilot manifold, hydraulic junction boxes, pressure gauges, and electrical junction boxes required to operate the BOP components. The nitrogen precharge pressure in accumulator bottles is 75. ______________________ for 3,000-psi (21,000-kPa) working pressure systems and 76. ______________________ for 5,000-psi (35,000-kPa) working systems. The 77. ______________________ carries hydraulic fluid to hydraulic junction boxes mounted on hose storage reels. As a minimum requirement, closing units of subsea installations should be equipped with accumulator bottles with sufficient volumetric capacity to provide the usable fluid volume (with pumps inoperative) to open and close the ram preventers and one annular preventer and retain a 78. ______________________ reserve. To overcome delays, 79. ______________________ are used in water depths greater than 5,000 ft (1,500 metres). The primary difference between a straight hydraulic system and an MUX system is that the MUX system 80. ______________________.

Answers to Review Questions

LESSONS IN ROTARY DRILLING

Unit V, Lesson 10: Marine Riser Systems and

1. F
2. T
3. T
4. F
5. T
6. F
7. F
8. T
9. F
10. T
11. e
12. c
13. b
14. d
15. a
16. b
17. e
18. d
19. c
20. d
21. mud booster line
22. riser-mounted conduit lines
23. riser handling tools
24. mechanical riser handling tool
25. hydraulic riser handling tool
26. buoyant modules
27. LMRP
28. ball joint
29. flex joint (answers to 28 and 29 may be given in either order)
30. flexible piping
31. T
32. F
33. T
34. F
35. F
36. T
37. T
38. F
39. F
40. T
41. g
42. f
43. h
44. i
45. d
46. b
47. a
48. j
49. e
50. c
51. b
52. a
53. a
54. b
55. c
56. c
57. c
58. b
59. c
60. c
61. T
62. T
63. T

64. F
65. T
66. F
67. F
68. F
69. F
70. F
71. hydraulic fluid
72. straight hydraulic
73. multiplexed (anwers to 72 and 73 may be in either order)
74. accumulator unit
75. 1,000 psi (7,000 kPa)
76. 1,500 psi (10,500 kPa)
77. hose bundle
78. 50%
79. multiplex electronic control (MUX) systems
80. provides the pilot signal very quickly